Secrets of a Celtic Mystic

Sacred Earth Prophecy

Also by Catriona MacGregor

Partnering with Nature: The Wild Path to Reconnecting with the Earth

Secrets of a Celtic Mystic

Sacred Earth Prophecy

Catriona MacGregor

Secrets of a Celtic Mystic: Sacred Earth Prophecy

Published by Catriona MacGregor
www.catrionamacgregor.org

Cover Photo "Sacred Bee" by Catriona MacGregor dated 1992 from traditional film

Printed in the United States of America

First Edition
Printed in the United States of America
ISBN: 978-0-578-84617-0 (print)
ISBN: 978-0-578-84618-7 (ebook)

For My Father, Joseph

Contents

SECRETS *of a* CELTIC MYSTIC

Sacred Earth Prophecy

Acknowledgments

I am so grateful to the many people who have helped to birth this book. This includes my many mentors, in human, animal and tree form, as well as the individuals who entrusted me to lead them on a Nature or Vision Quest. I am also most grateful to the many environmentalists who I worked side by side with to save the Nature we love. I owe a deal of thanks to all those who read the draft book in all its different stages.

I especially want to thank Denise Linn, a dear friend and mentor who has helped me" find and walk a path with a heart", and her daughter Meadow for offering a writing seminar at just the right moment. Eight years have passed since I first set"pen to paper" and as always the Creator in her/ his divine wisdom has set the time for its creation.

I would like to thank Barbara Kulle, who kindly read the manuscript and helped its progress with simple but profound words that helped me focus in no small way."This is who you are." she reminded me. Her words continued to light the path as I journeyed through the writing process.

I owe a debt of gratitude to Susie Herrick, for fully diving into the manuscript to offer concrete and detailed guidance of a reader's perspective on the material presented. I would like to thank my fellow MacGregor, Rob MacGregor and his wife Tish for also having a look and providing sound advice and encouragement.

My dowsing friend and talented energy worker Joan Rose Staffen, helped me to put my life in perspective that has come through in the writing.

I want to thank Dawson Church for his generous support and for having faith in my writing early on.

Rita Bjork and Rosaleen Bertolino were two people that had a look at the very first draft of the book many years ago. Their input from long ago has been most helpful.

Walker Laughlin lent a hand to help in editing and sorting out the many footnotes, and Sue Lascelles provided editorial expertise and advice.

I also want to thank Judith Kendra, the former Publishing Director at Ebury Imprint Rider books for her encouragement.

Last but not least, Tony Bonds, of Golden Ratio Book Design has brought a great level of professionalism and care to the book's creation.

Introduction

May I walk every day unceasing on the banks of my water, may my soul rest on the branches of the trees which I have planted, may I refresh myself in the shadow of my sycamore.

Egyptian Tomb Inscription, circa 1400 BCE

When we sicken the Earth, we ourselves become sick. This has never been more evident than during the 2019-2020 outbreak of the virus Covid-19 pandemic. Disturbed and polluted habitats, sick animals, and a climate-changed Earth attract pests and diseases. This was one of the first lessons I learned as a scientist developing habitat management plans in New Hampshire and New York State. It is a truth I experience as a scientist and leader working to protect trees, forests, and animals globally. Our world is out of balance, which encourages new diseases like the Covid 19 virus that has been unleashed upon the Earth.

Thus, our loss of connection to Nature has implications not just for the Earth, but ultimately for the human race. Will our descendants be able to breathe clean air, drink clean water, and see a natural blue sky? How will they experience the wonder and grandeur that wilderness offers if it ceases to exist? What will their lives be like if they never hear a live bird sing, touch a leaf of a native tree, or see a wild mustang run like the wind?

Sacred Earth warns of the dangers of "escaping" the natural world into an artificial one or solving planetary problems through technology alone. For example, plans to solve global warming by blocking the sun, or spraying materials into the atmosphere for the same purpose, will cause more problems, not less. Humankind is in this precarious situation now because we have lost touch with the Earth and natural forces. Besides, internet technology, computers, and AI present new problems, especially to generations born in the new millennium. While these technologies are powerful tools that help us in many ways, they are rapidly altering our lives and distracting us with a virtual world. Meanwhile, our real-world is slipping away. No band-aid technological solutions or "quick fixes" will save us. Deep learning about our planet and being guided by her ways is essential if we are to survive and thrive.

Modern socio-political "norms" and economic policies have led to our troubled times. They tend towards valuing material things over people and not valuing nature at all. This has led to the "over-domestication" of human beings, the perversion of our sacred Nature, as well as the destruction of our home, planet Earth.

Industrialization promised us a safer world and a life of greater ease and comfort, yet it took away the most important things that human beings could want or have: the freedom to spend their time as they wished. This is because many "industrial jobs" brought people in from the outdoors to work in factories. In addition, it created a "mono-culture" of work and vocation. Expecting human beings to work for long hours indoors instead of achieving their individual dreams, or access their creative skills and gifts leads to social and spiritual impoverishment.

While societal paradigms may limit and "over-domesticate" us, wild Nature can guide us back to wholeness. *Sacred Earth* explores what changes must occur to save humanity and the world from the onslaughts of diseases, food shortages, extreme weather events, pollution, and wars. We must reconsider our "role in Nature's sphere" and our relationship to all life. Humankind is not alone on planet Earth. Nature can support our *Wellbeing* as we *partner with her* instead of using her as a resource to use on our terms only. By adopting a new societal paradigm, that we are here to give more than to take, we will become the "Wise Humans" and the "Real People," we are meant to be.

I bring a unique perspective as a scientist, environmental journalist, and descendant of an ancient indigenous people, the Picts of Scotland.[1] Yet, from working with thousands of people from many countries, I have observed first-hand how all people have an innate ability to connect deeply with the natural world. I lead Vision Quests and found Nature Quest 20 years ago is to help people leave the limiting societal construct behind and tap into the wondrous natural and cosmic forces that surround us. By doing so, we will discover that as we heal the Earth, we heal ourselves. This connection is regenerative and can be profoundly life-changing. The fact that all of us have a predilection for being able to learn directly from plants, for example, was made evidently clear when I led a presentation at a large dowsing convention several years ago.

Before the convention, I carefully collected mistletoe leaves from an oak tree and made sure that every person in the room had at least one leaf to hold. After bringing people into a heightened state of awareness, I asked them to tell me about the plant that they held in their hands. While no one knew the name or type of plant from only its leaves, the vast majority felt that the plant was medicinal in nature. Some referred to it as "strongly therapeutic" and even "magical." Mistletoe has been used for centuries in traditional medicine to cure many conditions, including arthritis, headaches, and even seizures.

To the Celts and likely formerly the Picts, mistletoe was a sacred plant with both medicinal and magical qualities. Today, mistletoe is used in Europe as a treatment for cancer.[2] Many participants correctly "knew" the plant just by holding and tuning into the leaf held in their hands. This is a similar technique used by the great herbalist Dr. Edward Bach.[3] He could hold the petals or leaves of a plant in his hands and understand the plant's specific healing abilities. At the time, Dr. Bach had a thriving traditional medical practice and worked on vaccines and other remedies to combat diseases. However, after healing himself from a condition that he was supposed to die from in three months, using natural remedies, he became disenchanted with traditional medicine and discovered a "gentler" method to heal.

Bach intuitively understood that illness begins from unsettled states of mind, such as worry, fear, anxiety, and states of stress. Some plants could mend these emotional states and even prevent and cure illnesses. For example, the herb Mimulus helps remove fears of everyday life, such as being harmed by illness, accidents, loss, or misfortune. Dr. Bach's herbal remedies are available in many countries throughout the world today.

Thus, while none of the Dowsing Convention participants were trained botanists, the truth about the plant was revealed to most. It is possible to understand the world by listening deeply to the life forms we share the planet with. By doing so, we will find that there are natural solutions to many of humanities problems. For example, plants are masterful climate changers and could be our greatest allies in transforming the present and harmful climate trends.

Forests are sources of 65 percent of all clouds and 45 percent of all rainfall. The Intergovernmental Panel on Climate Change has identified natural aerosols emitted by plants as potentially one of the biggest game-changers for addressing global warming.[4] Plants and forests should be at the top of our list of climate healing heroes. We must stop cutting them down and start planting them!

My ancestors, the Picts, referred to trees as "The Ancestors of Humankind." While the idea of trees being our ancestors may seem far-fetched, it was the plants that created an oxygenated atmosphere. This new atmosphere resulted in the evolution of advanced species such as mammals and, eventually, humankind. We depend upon plants for our very survival because there would be no oxygen to breathe without them.

Sacred Earth is also a call to women to step up as authorities in their own lives. The connection between the debasement of Nature and the debasement of women was initially made by eco-feminists over a hundred years ago. In considering the treatment of women globally, we need to take a broad view, one that encompasses the divine feminine in all people, since men and women have within them both feminine and masculine qualities, just as cosmic forces contain both the Yang (active principle) and the Yin (receptive principle). Yet, these qualities are out of balance in too many individuals and societies worldwide, with the masculine or the "Yang" characteristics overpowering the feminine or "Yin" aspects.

Sacred Earth brings forward ancient knowledge and a future prophecy. The prophecy reveals that in these times of catastrophic shifts and "Great

Earth Changes," much will be taken away. The earth will no longer offer up her bounty as in the past, and many animal and tree species will leave; many already have. Ultimately humankind will survive—but in a very different world.

One day, our descendants will live in harmony with nature and lead freer, healthier, and more meaningful lives. We will succeed because individuals will evolve spiritually and abandon the socio-political and economic paradigms that have brought us to the brink. Many will have gained insight into and aligned themselves with what I refer to as the invisible, the wise and powerful force(s) that underlie and create manifest reality.

The Earth speaks to us all if we care to listen. Nature can show us how to free ourselves from the limiting and harmful constructs that society places upon the Earth and us. By partnering with Nature and the sacred forces that affect us, we can tap into our own divinity and "re-wild" ourselves. This will take a revolution in consciousness. It will happen, and it will be led by women and men who partner with Nature, see the sacred in all things, and honor "feminine" values of compassion, intuition, and safekeeping of life. In the end, we will liberate the Earth as we ourselves become free.

Preface

The world is too much with us; late and soon,
Getting and spending, we lay waste our powers; —
Little we see in Nature that is ours;
We have given our hearts away, a sordid boon!
This Sea that bares her bosom to the moon;
The winds that will be howling at all hours,
And are up-gathered now like sleeping flowers;
For this, for everything, we are out of tune;
It moves us not. Great God! I'd rather be
A Pagan suckled in a creed outworn;
So might I, standing on this pleasant lea,
Have glimpses that would make me less forlorn;
Have sight of Proteus rising from the sea;
Or hear old Triton blow his wreathèd horn.[1]

William Wordsworth

The wind dreams of freedom. It spreads its thought-songs over the Earth from the four corners of the world. The many waters—streams, rivers, glaciers, oceans, and lakes—carry those memory-songs in each sacred drop. These songs shaped the land, long before our barefooted ancestors left their five-toed mark upon the Earth. Then, only the stones, silent witnesses to the drift of continents, could hear them.

Later, when the plants, the trees, the animals, and the people came, these songs uplifted their souls. The songs taught them who they were and how to live. We once heard these songs, once danced to their rhythm. We lived in a world far more beautiful and wondrous. Yet through the ages, these songs were forgotten and faded away. Today, most of us no longer dance to those melodies.

We no longer know when the salmon will return to home streams to spawn. We cannot find the red fox's den hidden beneath green and yellow leaves. We do not know where to begin to look. We no longer sit upon the Earth with ease. Our bare feet miss the naked ground. This is because we turned away from the wilds, began calling the wild animals our enemies, or worse. We lost sight that trees and water are alive. We named them "resources" for our use as if they had no life of their own.

Perhaps you long for a world of wonder and mystery?

Have you ever stood barefoot upon the warm Earth filled with the songs of the sun? Or kneeled with your palms pressed upon the bare soil to smell the fragrant wild grasses? Or gazed upwards at the night sky, beckoning the stars' light to shine on your face? Have you ever walked quietly off the walkway through the woods to follow the deer path? Or come upon one and see its soul shining out of dark eyes?

If you have experienced one of these things deeply, you may know that Nature offers a doorway to a timeless realm. Beyond the door, epochal flashes light the path of life, as if to say, "This is Truth, this is Freedom, remember." The Wilds are within you as well as without. The whisper of the wind, the thundering hooves of a million buffalo, and even the lone cry of a wolf are sleeping inside of you. Awake! Nature can heal your soul.

Are you willing to step out from the constraints of a human society that calls itself "civilized" and "modern" and yet acts in ways antithetical to life? What will it take for you—for us—to protect and cherish each other, and what is left of life on Earth? If we have the capacity to destroy all life on the planet, we have the responsibility to ensure the protection and continuation of that life. Will we rise to the challenge?

We are now going through the birthing pains of this new age that the Mayans hinted at thousands of years ago. Our beliefs about other species—now considered "below us"—will undergo a drastic shift. No longer will humankind sit far above other life as the pinnacle of the greatest that life can aspire to. Instead, we will measure ourselves in relation to other species on an equal footing, with neither them nor us being the highest, or the lowest, of God's creatures.

From this place of humbleness, we will clearly see all our relations, the animals and plants, and understand them. They are our brothers and sisters, and they have great spirits that can help guide us. This earth island, floating in the vast sea of the Universe, is our very own living Gaia—the Gaia that shaped us and, in her own way, "mothers" us. Meanwhile, vast and unknowable cosmic influences are at work.

Ultimately, saving the Earth must start with you and me. It will not begin at the level of leadership until we have redefined what it is to lead. It will instead begin with the changes within each of us, and our willingness to act now.

Enter the great "heart of wilderness" within and without you. Let it teach you. Let it sing its songs to you. Let it fill you up so you can dance to those songs. Then, these will be added to you: the soul of a tree, the song of the whale, the wisdom of coyote, the wingbeat of the eagle.

This is the journey that I invite you on.

I. AIR

1. The Flight

It...recognizes that each item taught must be presented on the basis of previously learned knowledge and skills. It is designed so academic ground study units can be integrated with flight study units.

Cessna Private Pilot Training Syllabus, Introduction

The loud, whirring roar of the airplane's single-engine rose in my consciousness. As I glanced towards the horizon, the tangerine sky filled me with electric anticipation for the journey ahead. My hands lay on the yoke, as the golden dawn filled the cabin of the aircraft and settled gently on the tops of my fingers. I was piloting a single-engine two-seater Katana airplane on my way to the border of Mexico.

As the Executive Director of Texas Audubon, I would be visiting Audubon's Texas coastal sanctuaries and Sabal Palm Sanctuary, located on the Rio Grande River. Sabal Palm Sanctuary contains one of the last native Sabal palm forests. At 557 acres, it is no larger than a postage stamp when viewed from the air. Nevertheless, compared to much larger protected areas on the Earth, the sanctuary is home to many fascinating creatures, such as the chachalaca (a large flightless bird), the stunning emerald green jay, and the whimsical scissor-tailed flycatcher.

I had learned how to fly to see beyond what one might see on foot or in a car. This included miles of coastline and the wondrous birds and animals living where the sea meets the land. After spending months traveling by road, unable to access protected areas, I realized that there was really only one way to properly observe a state as big as Texas and its wild inhabitants.

Today, I would be flying from Austin to Brownsville to meet with the coastal Audubon wardens. These wardens dedicate themselves to protecting the birds' nesting islands by keeping predators and people away during nesting season and enriching the habitat by planting native plants. Lovers of the sea and sand, and avid naturalists, the five coastal wardens in Texas are the "salt of the earth." Perhaps no one exemplified a coastal warden's role better than Chester Smith; a former oil refinery worker turned animal lover and conservation steward.

I began working with Chester when he was a wiry, athletic 76-year-old who referred to the pelicans on Sundown Island as "my birds." I used to go with Chester on his boat to visit "his" island, where we observed the nesting pelicans and their offspring and discussed strategies for protecting them. Chester always kept a watchful eye on Sundown and would make sure that people stayed off the island during the nesting season. On one visit, as we stood on the island together looking out at the water, quickly looked down, pointed to the ground with a frown, and said, "fire ants." Fire ants are real pests and a hazard to nesting birds and their offspring and can bite and even kill young chicks. He explained how he was working to keep their population under control. Ever vigilant, I was very impressed with Chester's dedication. He was using his one very old and rickety motorboat to carry out his duties, so I made sure we got him a new one.

The Gulf Coast sanctuaries are among the largest continuous sanctuaries in the world. These islands provide critical habitat for many species, especially colonial water birds like the brown pelican. Green Island, along with the aptly named North Bird Island and South Bird Island, are thriving nesting sites for bird colonies. They were placed in the Audubon Society's care over one hundred years ago. The coastal islands stretch as far north as Galveston and as far south as the Mexican border. The islands cover 600 miles of shoreline, filled with islands, bays, and river mouths.

Green Island is one of the oldest bird sanctuaries in the United States. In 1923, it became one of America's first protected bird sanctuaries. Green Island is situated at the mouth of the Arroyo Colorado and close to the bountiful waters of the Laguna Madre, a superb breeding ground for egrets, other colonial waterbirds, and numerous other wading bird species. The tiny island harbors the largest colony of reddish egrets and roseate spoonbills on the continent. With their feathers shimmering bright fuchsia, these birds would be at the peak of their nesting season. Our annual survey would help us to compare twenty years of previous data. The survey would help us understand how the birds were faring despite predators, human disturbances, and pollution. It would also help us to develop an annual and five-year conservation strategy.

This year, thousands of colorful colonial waterbirds would be tending to their young on the protected islands, including snowy egrets, black-crowned night herons, roseate spoonbills, and reddish egrets. Like good tenement dwellers, they make their nests close to each other, either in hollowed-out depressions in the ground (as with white pelicans) or in low-lying shrubs (as in the case of roseate spoonbills). A cacophony of sound from adults and offspring then fills the air with a mighty din that would put a rock band to shame.

That raucous sound almost came to an end. Starting in 1900, Europeans and Americans took to a new fashion craze of wearing feathers, bird wings, and entire birds on their heads as hats. A single 1892 order of feathers by a London dealer (either a plumassier or a milliner) included 6,000 bird of paradise, 40,000 hummingbirds, and 360,000 various East Indian bird feathers.[1] In 1902, an auction in London sold 1,608 30-ounce packages of heron plumes. Each ounce of plume required the use of four herons; therefore, each package used the plumes of 120 herons.

The first Audubon Society was founded by two young Galveston women, Estelle Hertford and Cecile Seixas. Shocked at the great multitude of dead birds shipping out of Galveston harbor to ports of call, they founded the Society in 1899. At the time, they tried to stop the trade of feathers by making the wearing of feathers "unseemly." Hertford and Seixas raised awareness about the thousands of birds killed on their nests, leaving their babies to die unattended.

The decade of the 1890s was a ripening time for the women's movement. Though marriage and motherhood were still par for the course, and married women lived restricted lives expecting to cater to their families and households, young, single women like Estelle and Cecile had more options. While the women's movement had started earlier, the 1890s saw a big surge in women joining. Two of the largest women's suffrage organizations merged in 1890 to create the National American Woman Suffrage Association led by Susan B. Anthony.[2] Thus, Estelle and Cecille's entry into the world of "advocacy" was pioneering but certainly carried along by others in these times. However, little could they have dreamed that the organization they launched would survive them by over one hundred years. The birds that they sought to protect are still around today but in much fewer numbers.

In 1973, only 12 breeding pairs of brown pelicans lived along the Gulf Coast. In large part, their numbers plummeted from Dichlorodiphenyltrichloroethane (DDT). DDT is a colorless, tasteless, and odorless chemical used by the military in World War II. The military released it as a "product" to be applied in agriculture and as a household pesticide. This single chemical almost led to the extinction of the bald eagle and the peregrine falcon until it was banned in 1972.[3]

The banning of this harmful chemical was a huge step toward preserving the pelican population, but not an absolute answer to stopping their decline. By the time Chester took up being a warden for Sundown Island in 1986, there were only ten nesting pairs on the island.

Pelicans faced other threats along the Texas coast, including human traffic that disturbed their nesting island and ongoing dredging by the Army Corps of Engineers. The Texas coast is an important waterway for oil tankers, and the shallow waters are constantly dredged to allow access to ships. In the process, existing small islands might be destroyed, while dredged materials might be dumped to create new ones.

One afternoon while surveying Sundown Island, Chester and I discussed how one of the major impediments to safely raising pelicans to maturity was the constant dredging operations in the Matagorda Channel, near the nesting sites. Unfortunately, the Army Corps of

Engineers often dredged in the spring, which is peak breeding season for birds and most other species. Being a former oil company employee, Chester was aware that there was no chance that the dredging needed to move oil ships along the coast could be stopped. "Perhaps we can just ask them to change their schedule and not dredge in the spring?" I suggested. Chester initially shrugged, but a gleam in his eyes told me he thought we might have a chance. While we knew that there was no way that the Army Corps would stop dredging, perhaps they would at least stop dredging in the spring.

In fact, if we humans ceased taking actions that negatively affect wildlife during nesting season—such as the cutting trees in the spring when most mammals and birds are raising young—we would help species survive in no small way. Fortunately, since it would not affect their fiscal year budget, the Corps agreed not to dredge during peak nesting season. This was a significant step that led to the pelicans' dramatic comeback, along with the warden's conservation efforts. The following year, the population of successful nests and offspring significantly increased as a result of no spring dredging, and our other conservation activities. The following year, our Audubon coastal program received a Blue Medal in Conservation from the Governor.

This was a huge success for Chester and his birds. While Chester died in June 2011 after 25 years of service as a warden. He went on his last visit to Sundown in May when he was 90 years old to help lead the annual bird census. There were 2,029 nesting pairs of brown pelicans that year!

~

Back on that beautiful spring morning in 1997, I had started my journey to the airport at four in the morning, long before the sun came up. Before leaving our house in the Austin suburbs, I looked over at my seven-year-old son Joseph sleeping in his bed. I blew him a kiss and whispered, "*I love you as big as the sky.*" In the kitchen, I hugged my husband Paul as he leaned against the counter. He grinned tiredly, with a slight shadow on his chin giving him a rugged appearance, and his plaid green robe was still missing its tie. Holding a warm cup of coffee in his hands, he said quietly with a smile, "*I love you, so come back.*"

Driving twenty miles to the outskirts of Austin, I arrived at the small airport with small to medium sized planes and jets parked by the runway. Inside, I sat in the pilot's room to complete the flight plan for the day, handed it to the airport manager, and walked outside onto the tarmac.

The dawn air felt cool on my face as a golden-orange sky began to fill the horizon. "*I will be flying towards that glory,*" I whispered to myself. As I walked towards the single-engine Katana airplane, I smiled. The Katana is a highly maneuverable small plane with a large, curved windshield that provides a bird's-eye view of the land below and the sky above. It's perfect for surveying bird populations and identifying oil spills and algae blooms from on high.

I examined the plane's exterior and its internal equipment and dials to make sure everything was in working order before takeoff. When done, I opened the small metal door, climbed inside the cockpit, and sat in the pilot's seat. As I put on headphones to stay in touch with flight control and turned the small key in the ignition, the engine came alive with a roar. As the engine started, it thrust forward abruptly but slightly tipping its nose, like a racehorse straining in preparation for a race. Once the plane steadied, I slowly released the brakes and taxied towards the runway.

Flying is more magical than mechanical. Everything that matters about flight has to do with three things: the shape and surface of a plane, the flow of air around that surface, and horsepower. The Wright brothers knew this well. They understood that flying has as much to do with airflow and wing shape as engine power. While most other flying enthusiasts focused mainly on engine power and mechanical thrust, the Wright brothers explored the invisible secrets of the air.[4]

Within the Earth's atmosphere, everything is subject to air pressure. The animals, the trees, and human beings are so accustomed to this, like fish in water, that we simply overlook this unseen aspect of our world.

Yet, the flight would not be possible without it. An airfoil shape, which looks like a falling waterdrop cut in half, is the best shape for flight. The Wright brothers tested different wing designs before learning the secrets of the vagaries of air. When the shape of the wing is fashioned just so, magic happens. That magic is "lift." Lift occurs when the wing's shape causes the air above it to move more quickly than the air below it. When this happens, the plane is literally "lifted" into the sky by a spiral of air equal to the plane's wingspan.

The existence of "lift" is more than just how a plane, weighing thousands of pounds, can fly. Rather, it is a reminder that it is largely the invisible forces in our lives that can matter the most. We may believe that all that exists is what we can see and touch, but we would not have lives worth living without invisible forces. For example, emotions like love, compassion, kindness, appreciation, and forgiveness may be invisible to our eyes, but we are nothing more than walking automatons without them.

Flying has some practical and somewhat more subtle benefits as well. There are things that one simply cannot comprehend from a ground view. They say that when the Apollo mission sent back the first photograph of Earth from space, humankind's consciousness took a giant leap forward. We saw the Earth as a living jewel suspended in the vast silence of space for the first time. Some saw Earth as a sentient being of great beauty as well as great vulnerability. Scientist, James Lovelock, developed the Gaia Hypothesis: that Gaia creates the conditions (such as weather) and the chemistry that support its own life and the life of all its inhabitants.[5] This was a bold vision. It suggests that humans are just one of the multitudes of life forms supported by a conscious being, the mother to all earthly life.

Like the Wright brothers, I learned at a young age that there is more to this world than what we can usually see with our eyes and feel with our hands. For example, although invisible to many, I could see luminous, radiant "clouds" of light around people's bodies. These light clouds that generally appeared 2-3 feet away from the physical body were oval-shaped and beautiful. I was to learn later that these luminous clouds of colors are known as auras. Auras are sometimes referred to as "energy fields," and they have even been photographed and recorded with special equipment. While auras have been reflected in artwork going back thousands of years, Nicola Tesla was the first scientist to take a photo of an aura in 1891.[6]

I understand auras as reflections of a person's thoughts, feelings, and emotions. The long-standing colors in a person's aura reflect long-held emotions and beliefs and even personality traits. Brief fluctuating emotions can also fleetingly appear in an aura at times. Witnessing a person's aura always conveys information about their emotional state, health. and personality style. Much of this information is conveyed at a visceral level of understanding and difficult to put into words. However, over the years, I have understood the meaning of some of the aura colors. For example, dark blue is associated with one's power of will. Green is associated with action and, depending upon the hue, reflects either physical skills or mental acuity. Yellow and gold are associated with understanding, and a holistic way of knowing, a pearl of inner wisdom if you will. Rose reflects a loving attitude, and lavender or violet reflects intuition and connection to spiritual dimensions. While these are just broad generalizations, and there are much finer and more complex details to understand, the existence of these "subtle dimensions" prove that we are much more than our physical bodies.

I could also sense things about the Earth and Nature that others seemed to miss. There are auras around animals and trees, although these appear differently than in people. This "sight" led me to know and experience that the manifest world is only a secondary effect of something deeper and more enduring. In fact, auras link us to an aspect of ourselves that is transcendent and survives the physical body. The physical world is merely a reflection of the vibratory dance supporting all of creation.

Look inside an atom, and you will find nothing but tiny, oscillating fields in what appears to be empty space. Yet we, with our atoms, can "plant" our feet firmly on the ground and feel supported by the Earth. The only thing stopping us from falling into a seeming abyss is the simple fact that our atoms are oscillating or vibrating at a different frequency than the ground. Oddly, while you and I contain the unique dancing "fields" within every cell in your body, we are never dancing alone. Our auras have a series of cone-shaped spirals along their boundaries that allow emotional energy to flow into and out of the larger energy fields around us. We live in a boundless ocean of coupled energy; there is no you, no them, no me, just "all of us" together. This interconnectedness extends to all living beings, including our animal and plant brothers and sisters. Our ability to moderate a good flow of beneficial energies between ourselves and the world around us is essential for good health and vitality.

Over the years, I have realized that the life journey that we lead in the manifest world resides alongside this deep, enduring, and subtle reality. When we recognize and heed the "calling" of this nonmaterial realm, we step into a pearl of wisdom, a knowing, and a truth. A wisdom that can guide us to live a life filled with meaning, magic, and mystery. Then we are no longer simply subject to the vagaries of the profane but uplifted to experience the "more" of existence: more beauty, more love, more wisdom. It is both an extremely humbling, freeing, and glorious experience. It is like a plane that has lift.

~

Sitting in the pilot's seat at the Austin airport, the runway's pale-green directional lights glowed in the gentle dawn of the new day. I placed my foot firmly on the brake and opened the throttle. This warmed the engine for the flight ahead. In a loud, crackling command, the air traffic control radioed the okay for takeoff. My foot firmly on the throttle, the plane engine engaged with a roar, and we were off. As the plane accelerated down the runway, a strong crosswind made it bob slightly from side to side. As the plane picked up speed in the last few feet of runway, aerodynamic forces took over. We rose into the sky as if lifted by the invisible hands of a beneficent being. Takeoff is my favorite part of flying. It is the moment when a heavy metal plane is transformed into a wondrous object of flight; skywards, we rose.

As the sun appeared near the horizon, shining golden pink light into the cabin, the Katana and I rose above the pull of Earth's gravity and into the "arms" of "lift." Rising effortlessly into the air, I knew I owed my deliverance, in part, to the invisible forces of the sky.

2. How the Animals Shape Us

If men would pay more attention to these [animal, plant] preferences and seek what is best to do in order to make themselves worthy of that toward which they are so attracted, they might have dreams which would purify their lives. Let a man decide upon his favorite animal and make a study of it, learning its innocent ways. Let him learn to understand its sounds and motions. The animals want to communicate with man, and man must do the greater part in securing an understanding.[1]

Brave Buffalo, Chief of the Teton Sioux

A good pilot routinely looks at the lighted dials of the panel instruments on the airplane dashboard during flight, as well as landmarks below, to make sure that she is in the right place at the right time. While the Piper has a wide array of instruments to orient a pilot as to direction, speed, and altitude, migrating birds travel thousands of miles without any of these. Many of them even fly at night.

Birds accomplish their amazing migrations by following landscape features, geomorphological landmarks, air mass differences in temperature and moisture, the stars, and Earth's magnetic fields.[2] With these extrasensory skills, they find their way. Their behavior reflects an innate intelligence that has guided the development of their kind through the eons.

Two-thirds of the birds found in North America migrate every year along the same sky routes.[3] The Texas Gulf Coast is one of three of the most-used migratory routes. Known as the Central Flyway, the coastal route is a super-flyway traversed by billions of birds annually. This was the same route that I was following in the Katana.

In the spring, birds fly north when the warm weather allows insects, fruits, and many of their other food sources to become plentiful. To care for their young, nesting birds take advantage of the extra daylight hours that northern latitudes bring. In the winter, many of the birds that nest in Canada and the United States fly far south, to the West Indies and South and Central America, to escape the harsh weather. Birds that travel these distances are called neotropical migrants. Astonishingly, one of the longest migrations, a 44,000 miles journey, is accomplished by the four-ounce arctic tern.[4] Terns can live up to 30 years, which means that long-lived individuals may rack up the equivalent miles traveled for three trips to the moon and back.

What if we could learn from the birds that migrate thousands of miles to read the signs, invisible to humankind, that help them flawlessly navigate the Earth? What if we could see with our own eyes, beyond the visible spectrum of light, into the ultraviolet or infrared ranges? Not only would we be able to navigate from on high like the birds, but we would see other invisible patterns. Like the intricate patterns on flowers that we humans never get a glimpse of, but that insects like bees plainly see. Yet, we can perceive the energetic life force around living things. We are born with this skill, but forget it, as we are raised to focus solely on our world's material aspects. Yet, by reconnecting with nature, learning from her, we can open a window to our more subtle perceptions and intuitive knowing.

We are meant to connect with animals and plants, not superficially, but at the deepest levels. We need to learn things from Nature that no human being can teach. These ways of knowing are now needed more than ever to navigate the changes sweeping our planet safely. What if we could hear, with our own ears, the long-distance call of an elephant at a range so low that we normally cannot hear it?[5] The truth is that humans perceive but a limited amount of information. We go through life with blinders on, missing an extraordinary amount that surrounds us. The animals can help us to experience the world in ways previously beyond our ken.

One of the oldest poems in English, "The Seafarer," references how ancient sailors followed the lead of one of the greatest marine animals of all, the whale:

Over the whale's path,
Widely wandering
All earth's corners.
Comes oft to me
Greedy and eager,
Lone-flyer screeching
Whets for the whale-road
The heart unwearied,
Over the sea's hold.[6]

As this poem shows, animals have served as guides and helpers to humankind for eons. They can help us survive physically, yet they do much more than that. They can inspire and uplift us spiritually. This is what Saint Francis of Assisi, citing Job 12:7, meant when he said, "*But ask the animals, and they will teach you; the birds of the air, and they will tell you; ask the plants of the earth, and they will teach you, and the fish of the sea will declare to you.*"[7] Animals (and plants) are intelligent beings that can open doors to different ways of knowing. Today scientists are catching up to what the ancients have always known.

In the past, the human-animal relationship was revered as a mutualistic one that reflected a sacred union. The animals, and the trees, helped to shape the evolutionary development of our consciousness. Teaching us the way of the wilds, they speak to us in a way that no human being can. Native peoples refer to animals as "brothers and sisters," evincing a familial bond. Beyond their physical relationships with the animals, the trees, and the earth, they believe in a deeper, invisible connection. A connection that aligns them with what they refer to as spirit animals,

animals whose energy and spirit resonate with a specific person, tribe, family, or group of people.

Perhaps the most important benefit of working with animals on a shamanic or mystical level is that we can access knowledge beyond the ken of any single person or society. This can be very valuable in times of great changes, as in the case of Chief Plenty Coups of the Crow Nation. Chief Plenty Coups had a prophetic vision in 1850 that guided the tribe in the face of the onslaught of the Europeans to the Americas.

The Crow people, along with other Native American tribes, were dealing with events and people that they had never encountered before. While they were familiar with hardships, the influx of these new, "strange" people from across the sea presented them with a perplexing situation. For example, the Europeans claimed "ownership" of land, a concept foreign to the native peoples. The newcomers were also bringing unusual objects like metal pots and guns that had many useful purposes. Meanwhile, many thousands of native people were dying of new diseases brought by the newcomers, and some were even being hunted down and killed. Because of this, members of the tribe wanted to go to war with the Europeans, others wanted to trade with them, and others simply wanted to avoid them altogether. For everyone, it was hard to know what to do with these new threats. It was at this critical time that Chief Plenty Coups took a vision quest. He was just nine years old.

During the many days and nights of the quest, he fasted and then received a powerful prophecy:

Everywhere I looked, great herds of buffalo were going in every direction, and still others without number were pouring out of the hole in the ground to travel on the wide plains. When at last they ceased coming out of the hole in the ground, all were gone, all! There was no one in sight anywhere, even out on the plains.[8]

After the buffalo had vanished, Plenty Coups then saw that,

Out of the hole in the ground came bulls and cows and calves past counting. ...many were spotted... And the bulls bellowed differently too, not deep and far sounding like the bulls of the buffalo but sharper and yet weaker in my ears...They were not buffalo. These were strange animals from another world.

Then, Plenty Coups saw

only a dark forest. A fierce storm was coming fast. The sky was black with streaks of mad color through it. I saw the four winds gathering to strike the forests and held my breath. Pity was hot in my heart for the beautiful trees. I felt pity for all the things that lived in that forest, but was powerless to stand with them against the Four Winds that together were making war... [I saw the] beautiful trees twist like blades of grass and fall in tangled piles where the forest had been ... Only one tree, tall and straight, was left standing where the great forest had stood ... Standing there alone among its dead tribesmen... 'What does this mean?' I whispered in my dream.

Plenty-coups then heard a voice:

In that tree is the lodge of the Chickadee. He is least in strength but strongest of mind among his kind. He is willing to work for wisdom. The Chickadee person is a good listener. Nothing escapes his ears, which he has sharpened by constant use... He gains success and avoids failure by learning how others succeeded or failed, and without great trouble to himself... "Develop your body, but do not neglect your mind, Plenty-Coups. It is the mind that leads a man to power, not the strength of the body."[9]

The prophecy revealed that unimaginable changes were coming to their homeland that would forever shift their lives and the lives of their descendants. Under a decade, millions of buffalo would vanish from the plains, shamefully shot by the "pioneers" seeking to rule the American West. Eventually, the domesticated cow would take the place of the mighty buffalo. The Four Winds represented the Europeans and their sweeping subjugation of the American continent. The forest of trees represented the many tribes of the Plains Indians, all of which would fall to the white man. The one tree left standing represented the Crow people or, as they called themselves, the "Absarokees." They alone could be left "standing" if they watched and learned like the chickadee instead of resorting to force.

Following Plenty-Coups' prophecy, the Crow Nation is the one tribe of the plains that chose not to make war against the European immigrants.[10] In turn, they are among the few Native American nations that retained a portion of their ancestral homeland, like the Chickadee perched on the single remaining tree. The chickadee served as a spirit animal, bringing Plenty-Coups and his people an important prophecy.

Today, we modern humans face even a more challenging time of upheaval and change. Climate change, new dangerous viruses, economic and political disruptions, and unrest are all symptoms of drastic adjustments to our way of life. Many are stepping to the fore with pure technological "solutions" out of touch with the natural ebb and life force flow. More than ever, what is needed is first to understand what the driving forces behind these dangerous shifts are and which solutions are most effective to slow or prevent them from continuing. We must reconnect with the Earth and adapt. To do so we need to listen and learn from Nature.

All animals, and plants for that matter, have an invisible divine or energetic reality beyond their physical existence. Just as human beings have auras, animals and trees also have energy that extends beyond their physical bodies. Human beings can and do interact with these fields of energy. Spirit animals have been referred to as "totem" animals by Native Americans and as "power" animals, a term especially prevalent in the 1960s and 1970s. I prefer to use the term "spirit animal," which more closely reflects how we deal with the invisible "over-soul" of the species.

Shamans, and mystics like myself, that work with spirit animals are routinely lifted from a limited personal and societal view of the world to access sacred visions, future prophecies, and different dimensions. The ancients too had the ability to align energetically, and certain members of their community had psychic ties to certain animals. These special "relationships" helped them avoid danger from predators and locate certain animals in the hunt. In some cases, game animals would come towards those they had a special tie to, pulled along by the magnetic quality of and connection to a person's or group's consciousness. Imagine, then, the jarring emptiness the Plains Indians must have felt when their buffalo were almost hunted to extinction. Not only was this a loss of a valuable food source, but the buffalo's disappearance would have been like a psychic death.

Spirit animals have guided humankind for thousands of years. Most cultures—including Egyptians, Native Americans, Hindus, Mayans, and others too numerous to mention here—recognized the importance of animal and plant spirits. For example, from my own heritage and Celtic culture, each clan has a "clan animal" in much the same way that the Native Americans had a totem animal for their tribe (such as the Crow People). The oak tree and the lion are two of the most important symbols for my family clan, the MacGregors. These symbols are found on our coat of arms and badges. It is believed that they conveyed specific traits that benefited clan members. Almost all Celtic gods and goddesses are linked to one or more spirit animals. In the case of Andraste, the unconquerable goddess worshipped by the great Celtic Druid and warrior queen Boudicca, the hare is the goddess's animal ally.[11] This is why brave-hearted Boudicca, with her long red hair flowing in the wind, released a hare before one of her greatest battles against the Romans.

Many may have forgotten, or have never been told, the story of the Buddha and his special relationship with lions. Before his incarnation as the Buddha, in his previous life, he was Prince Sattva, the son of King Maharatha. He was also far along on a spiritual path in that life, and while walking near the forest, he came upon a starving lioness and her hungry cubs. To save her life and those of her offspring, the Prince offered his body to the lioness to eat. Through his supreme act of generosity, he perfected all of the tenets he held closely, such as renunciation and equanimity, to be able to move to a higher level of evolution and incarnate in his next life as the Buddha.[12]

The Hindu God Vishnu has had ten primary incarnations before becoming Vishnu, including his lifetimes as a fish, a tortoise, a boar, and a lion, along with his previous human lives.[13] In other cultures, the gods themselves are represented by animals, as in the case of the Egyptian god Horus, who is sometimes represented as a falcon and at other times as a person with a falcon's head.[14] Horus was the god of the sky for the ancient Egyptians, one of Egypt's greatest and most powerful gods. Not only did the ancient Egyptians honor a variety of animals, like jackals, hawks, snakes, and cats, but many of their gods were entirely or partly comprised of animals.[15]

While these ancient stories and myths may seem out of touch with our modern lives, they point out that animals still influence us in subtle and not so subtle ways. For example, the tale of King Arthur, the chivalrous defender of England, makes this point. King Arthur is said to have turned into a raven at his death.[16] It is still considered unlucky today to kill a raven in parts of Wales and Cornwall. Today, six ravens reside at the Tower of London.[17] Treated like royalty, their very presence is linked to the belief that they help ensure the City of London's invincibility. If you visit the tower, you may hear the "caw caws" of the ravens Branwen, Hugine, Gwyllum, Thor, and Baldrick. When many today would scoff at the possibility that an animal could hold sway over human affairs, at the Tower of London, a Yeoman Raven Master whistles for his royal charges to come and feast upon their daily meal.

In *Mystery of The White Lions: Children of the Sun God,* Linda Tucker tells of her safari vacation to Timbavati, on the edge of South Africa's Kruger National Park.[18] One night, when a small group of tourists traveled deep into the bush to see if a white lion cub had been born, their open jeep broke down in the midst of an angry pride of lions. They were rescued by a Maria Khosa, a 'lion shaman' and a renowned *sangoma,* or medicine woman. That night Linda saw Maria walk safely and fearlessly to the jeep, passing within feet of the lions. Maria's presence had a calming effect on the entire situation, and they were able to get help to fix the jeep and eventually drive away. Maria explained later to Linda that the lions are not just physical beings but guardian spirits that she is deeply connected to.

Maria's ability to silently "communicate" with and relate to the lions is an example of the sort of relationship that a human being can have with a wild animal that many find hard to believe. Yet, we are told through folklore and ancient stories that there was a time when people could "talk" to the animals. This was most likely not via spoken language but through a type of "communion" with the animal at an energetic and spiritual level. This leads one to appreciate the suffering and jarring emptiness that the Plain's Indian's endured when their buffalo were almost hunted to extinction.

According to Carl Jung, just as each individual has its own unconscious and conscious mind, the human race collectively has what he called an "Oversoul."[19] The Oversoul is like a collective unconscious that helps individual members of a species organize their inner world and better understand the outer one. One aspect of this universal mind, according to Jung, was the concept of universal memory, a collective recollection that all people share. I believe that all species have an Oversoul specifically for their kind.

The Oversoul is then an "infinite field of knowing" accessible to every individual member of that species. This becomes obvious in the cases of when an animal's "instinct" guides it. For example, how does a newly hatched butterfly know that it now must eat nectar from flowers and no longer the leaves of plants, its main food when it was a caterpillar? How does a migrating bird know when and where to migrate? How do mother animals know how to care for their young? This knowledge and more are held within the Oversoul of that species. When we interact with a spirit animal, we are actually communicating with the Oversoul of its species.

3. Pelican

Kind Pelican, cleanse my filth with Thy blood, one drop of which can save The whole world from all its sin.[1]

Thomas Aquinas

The Piper flew onwards beneath small white clouds dancing slowly on the unending stage of the sky. Beneath the impartial eye of the blue heavens, the sensuous shape of the land calls to one. The Earth's beautiful body reveals streams, lakes, groves of trees, and rivers that curve and mimic a feminine form. Here the river bends gracefully around a pebble beach resting in its innermost curve, there the native trees reach lovingly towards the coyote bushes. Beyond the trees and the river, golden-hued grasslands stretch in languorous expansion towards the ocean.

Due to favorable winds, I flew in one sinuous arc over the Laguna Madre to look at colonial waterbirds nesting in the region. I hoped to spot the endangered pelicans nesting on or near the Audubon coastal islands. The long-lasting pelican has seen thousands if not millions of species come and go. Yet it was brought to the brink of extinction by our species, 30 million years it's junior. Pelicans existed in the Eocene and have remained largely unchanged as a species.[2] This makes them 29,800,000 years older than *Homo sapiens*. Pelicans came into being when the mighty Alps rose in Europe, and grasslands spread far upon the Earth, bringing grazing animals in their wake. Camels roamed the White River Badlands of America, undoubtedly grazing near tiny *Eohippus*, the horse's three-toed ancestor.

I imagined *Eohippus*, the first horse, trodding gently on soft, moist ground, leaving delicate three-toed hoof-prints spread like little fans upon the Earth. She would move seamlessly through the verdant ferns, large dark-green fronds hiding her slight build. Shaded from the sun and predators by dawn redwood trees, *Macginitiea* branches, and leaves of white cedar, elm, birch, and alder, she would have smelled the warm moist air, filled with the sweet richness of pollen and palm flowers. Arching her neck to reach down towards the new leaves of an ancient buttercup, she would give a little pull and then chew the leaves quietly, her lightly spotted coat blending within the dappled shadows.

Eohippus' vast tropical forests were unlike any known to humankind, stretching for seemingly unending miles in all directions.[3] Far to the east, there were enormous lakes, the size of inland seas. To the north, the sub-tropical forest gave way to temperate forests that stretched as far as the North Pole, covering that now ice-filled domain with green plants. This was the Eocene epoch, fifty-two million years ago. This age was the dawn of many new mammal species that, like *Eohippus*, thrived on the green plants reigned in the former Earth's sultry climate. Trees covered almost every square inch of the vast supercontinent Laurasia.

Only two feet long and eighteen inches high, about the size of a fox, *Eohippus* was quite diminutive compared to the large and powerful horses we know today.[4] Yet compared to most other Eocene mammals, it was of average size. *Eohippus* remained much the same for over 20 million years. It was perfectly adapted to its habitat and the Earth of that time. The verdant Eocene epoch, occurring millions of years before *Homo sapiens* even existed, might have been as close to what we consider Eden as the Earth has ever come.

North America, the birthplace of the horse, was ruled by plants. It would be a good fifty-one million and eight-hundred thousand years before those three-toed hooves of *Eohippus* would become the single hoof of *Equus*, the horse as we know it today. As the sub-tropical forests began to make way for miles of drier open grasslands, and the poles began to get colder, *Eohippus* also changed. The early horse evolved over several stages to cover vast distances of grasslands on powerful legs. Its speed and endurance made it a perfect match for the new habitat of open plains and immense prairies. Today, the horse's running prowess, of speeds of 35 miles per hour or more, rivals the wind.

~

Unlike the horse, the pelican has retained its size and shape over 30 million years.[5] Over the millennia, once *Homo sapiens* appeared on the Earth, pelicans made quite an impression. The ancient Egyptians revered the bird and associated it with the goddess of death and the afterlife.[6] The Egyptian Pelican goddess, Hemet, had the power to assure safe passage for mortal souls through the underworld. This is why pelicans are often depicted on the walls of Egyptian tombs. One such ancient engraving dates from as far back as the fifth century B.C.E.

In the Middle Ages in Europe, pelicans continued to play an important role. They were depicted in early Christian art and statuary.[7] The Pelican was compared to Christ on the Cross, who gave of himself for humanity's welfare. In Medieval religious art, the pelican is sometimes placed on top of the cross to represent resurrection and the risen Christ. Even the long-lived Queen Elizabeth I adopted the pelican as a symbol to portray herself as the self-sacrificing mother of the Church of England, giving of herself for her people's welfare.[8]

Saint Thomas Aquinas' hymn, "Lo, the Full Final Sacrifice," expounds upon the Pelican-Christ saving humankind:

O soft self-wounding Pelican!
Whose breast weeps Balm for wounded man?
That blood, whose least drops sovereign be
To wash my worlds of sins from me.
Come love! Come Lord! And that long day
For which I languish, come away.
When this dries soul those eyes shall see,
And drink the unsealed source of thee.
When Glory's sun faith's shades shall chase
And for thy veil gives me thy Face.[9]

Beyond the Piper's rounded windshield, the Laguna Madre, a narrow body of water, appeared like a turquoise bracelet of interlocking links upon the land. The Laguna Madre is an abundant estuary, home to thousands of species that feed in the rich shallow waters. Its seagrass meadows harbor more finfish than anywhere else on the coast. Because no rivers feed the Laguna Madre, it is the only lagoon in the United States that is saltier than the ocean. The Laguna's waters are filled with saline-loving seagrasses and surrounded by drought-tolerant plants such as mesquite trees, sweetgum, prickly pear cactus South Texas ambrosia.

Around the lagoon, a wide diversity of animal species flourish in the coastal savannah and brushlands. Within this tapestry of land, water, and plants, exotic animals abound, including the blue-winged teal, the nine-banded armadillo, Kemp's ridley sea turtle, the rare ocelot, and the jaguarondi. Both cats are smaller versions of and cousins to larger cats like pumas, cheetahs, and leopards. They are built for speed and survival in these arid lands. Yet each has a roar befitting even the greatest large cats.

At the crossroads between temperate and tropical climates, I imagined the Laguna Madre, millions of years after the Eocene and long before the Spaniards' coming, in the days it was cared for by the indigenous peoples. The lagoon's salty surface waters would be enlivened with the quicksilver movements of thousands of fish below its surface. Above, the air would be filled with hundreds of lyrical bird songs, some delicate and sweet to the ear, others loud and awakening, and others low and mournful. How much more alive the earth was when filled with those songs, each one with its own cadence, pitch, and rhythm! The sounds would be carried in gentle breezes that smelled salty and sweet, with wafts of fennel from the blooming yucca flowers.

In spring, some plants might have been covered in hundreds of moths like a living shroud. Inland, a mother ocelot, with a leopard-spotted coat and her two young tawny cubs, might have slinked towards the lagoon. At the same time, an enormous flock of bright-pink roseate spoonbills gathered at the water's edge, oblivious to any danger. During storms, strong offshore winds would have blown, bringing their invigorating blast of salt and seafoam. These winds would overshadow all other sounds, like the percussion section of an orchestra taking the lead. This natural symphony has not been heard for at least two hundred years.

The daydream came to an end as the plane passed directly over the Laguna Madre's northern end. Below, brown, scruffy land made way for dark-green waters in the shape of a sinewy snake entering the lagoon. Once there, these once-green waters turned brown and soupy. Since the 1990s, the Laguna Madre has experienced a persistent bloom of brown tide algae caused by pollution. That morning, the brown tide covered at least a third of the Laguna Madre's surface, marring its former pristine beauty. Looking down at the sickening brown tide, I knew its ill effect on colonial waterbirds, fish, and the many other animals that depend upon the lagoon.

Just three to five human generations before us, approximately 200 years ago, the great diversity and abundance of life was largely still intact. The human race, which originated 200,000 years ago, lived side by side with teeming Nature for most of its existence. If humankind's time on the planet were condensed to one 24-hour day, it is only in the last 1-2 minutes of that day that we have sent thousands of species into extinction.[10] Yet, even just five minutes ago on the same clock, the grandness of wilderness, the abundance of pure waters in lakes, rivers, and oceans, reigned on the Earth, along with a dizzying array of life. *Homo sapiens* is a species in its relative infancy on the Earth, compared to other life forms, yet we hold the fate of many other species in our hands.

In the 1800s, the great naturalist John James Audubon witnessed a flock of passenger pigeons a mile wide that took three days and nights to fly overhead until the very last straggler came into view.[11] Ornithologists today believe that there were likely over 3.5 billion birds in that one flock! Just one hundred and fifty years ago, there would be bird songs from every wooded grove filling the air with a sweet symphony. The music of rushing rivers, singing streams, and the whispering wind across the land would fill one's consciousness. The earthy rhythmic voices of frogs and crickets would enliven the night, punctuated by the owls' mysterious calls, and ending with the long crescendo of the wolf, silencing all the rest. Imagine the unbridled sounds of the Earth and her wild creatures.

Life thrived among biodiverse brilliance: crystal-blue skies with clouds reaching the heavens like mystic castles, great waterfalls draped in ethereal rainbows of light, herds of buffalo a million strong, rivers as impressive as the great Nile, lakes filled with water that shimmered with a wholesome purity, a sky filled with thousands of birds, oceans brimming with fish flashing silver scales, and forests of trees so grand they looked like immortal giants planted them. Trees like the long-lost American elm, whose acre-sized crown could offer shade for an entire town populace out for a picnic under its boughs.

What might it have been like to live in the days of the great diversity and abundance of life? In all of this world's aliveness, must not the human spirit have been more alive too? Imagine standing with James Audubon and watching that million-fold flock 200 years ago; or being a member of a Native American tribe 1000 years ago when 75 million buffalos moved like thunder over the plains; or even being an ancient Pict, standing 600 years ago on a cliff on Orkney, at the northernmost tip of Scotland, watching a pod of hundreds of whales sending up frothing fountains out of the sea? These would indeed be awe-inspiring sights to behold, and then there were the rivers.

Rivers once ran so wide, deep, and powerful that they rushed like major arteries of life on a thirsty land. The Rio Grande along the border of Texas and Mexico was indeed a "Great River." In the sixteenth century, the Rio Grande's mouth was more than 30 feet deep, allowing large ships to sail many miles up the river.[12] Large steamships carrying passengers and cargo over 100 miles upriver were recorded as late as 1907. It flowed wide, deep, and long, starting 2,000 miles from its birthplace in the Colorado Mountains. It journeyed south to Mexico, at which point its fresh waters melded into the salty waters of the ocean. Today, after hundreds of years of human activity, including drawing much of the Rio Grande's waters for agriculture and other uses, the river is a mere shadow of what it once was.

Indeed, there have been many earth changes over the eons, and never more than now, under the thumb of Man. For many millions of years, evolution occurred without us and carried on with the millions of species that arose over time. Today, humankind plays a domineering role in the continuing evolution of the planet and all life upon her. With great power comes great responsibility. Do we not owe other species that have played a key part in creating the world as we know it, respect?

~

Flying along the coast, I realized our team's efforts to help the pelicans survive was not just about pelicans, but also about us. The ancient Egyptians and even Queen Elizabeth could never have imagined the possibility that not just pelicans, but many other species, could and would be wiped out by the human race. What if the pelican no longer existed to fire the imagination of our children and our descendants? The pelican's duty as a saver of souls, and its virtuous link to selflessness, are significant. These same virtues are needed more today than ever before. The human world churns out leaders more interested in their own gain than performing selfless acts for others. I could not help but think that the ugly brown "stain" on the Laguna Madre was a visible symbol of society's failure to cherish the act of giving.

As the plane banked and tilted, specks of white appeared along the shore. I flew lower, still banking the Katana, while the specks grew larger. My heart jumped for joy. A large flock of at least 100 *Pelecanus erythrorhynchos*, or white pelicans, was basking in the sun on the coastal sands. The pelicans' shiny feathers exuded a subtle pink glow from the morning sun. Within moments a flurry of bright white exploded as the enormous flock spread their wings in unison and flew, spiraling towards the sun.

Their coordinated movements gave the appearance, from above, of a giant white silken scarf billowing gracefully over the face of the land. My heart filled with awe and tenderness for the world as I thought to myself, *there is still beauty yet.* It is a beauty and a wonder that our children can yet still witness. But for how long? Even just 30 years ago, the environmental problems seem minor compared to the global-sized ones that face us today.

My love for a natural world, vanishing before my eyes, has grown with the years, and so has my sorrow. I hope that someday we will take up the pelican duties, as selflessly envisaged by our predecessors, and turn our world around. I know that it will be the animals, and the Earth that will guide us then.

4. Eiocha

Once upon a time, there was no time. There were also no Gods and no man or woman to walk the land. There were only the depths of the sea and its dark, eternal quiet . . . where the sea met the land; a sea-horse came to be born of white sea-foam. She was a mare, and her name was Eiocha.[1]

The Oran Mór

The sun hung higher and brighter in the sky as the Piper slowly glided southeast toward the coast. I tilted my eyes downwards away from the sun's bright rays, adjusted the throttle, and began to slow the plane's airspeed. I flew in a wide, turning arc eastward over the long, narrow strip of land known as Padre Island (or "father island," being no doubt the consort to Laguna Madre, "mother lagoon"). Padre Island is a 113-mile-long barrier island located off the southern coast of Texas. Part of Padre Island is a nationally protected seashore with a great diversity of plant and animal life. Different coastal plants reside in the island's center, such as seashore saltgrass, bushy bluestem, sea oats, marsh hay, and cordgrass. Near the ocean waves, the island has stretches of bare sandy beaches.

It was still morning, but the sun was well above the horizon, with rays of light streaking across the waters of the gulf. As I looked out the Katana's windshield, movement along the sandy shoreline caught my attention. I took another wide arced turn over Padre Island to get a closer look. Halfway through the plane's turn, a lone horse and rider came into view on the beach below. The horse was white, with a long tail and mane moving like white fire in unison with its muscular movements. The horse and rider moved together rhythmically. As the horse's front legs stretched elegantly in a graceful dive to meet the ground, little brown clouds of sand flew up from the impact of its hooves. I arced the plane one more time at a lower elevation to get a closer look. As the horse came into view, its movements were entrancing. With the wings tilted and the plane turning slowly in a spiral, I kept the horse and rider in view for as long as possible. The sight brought me back to my time in France and the Camargue's stunning white horses, twenty years earlier.

Then, I noticed about a sixteenth of a mile ahead of the horse and rider, what looked like small white dots begin to move and flutter. It was a large flock of white pelicans. As the horse and rider approached, the birds startled and began to take to the air. As they became airborne, their opened wings hid the horse and rider. I strained to look beneath the turning flock, yet, the horse had completely disappeared under hundreds of ethereal ivory wings glinting in the sun. Although I lost sight of them, I imagined them flying above the sand, under a shimmering lake of light, an illusion created by the sun's reflection upon thousands of brilliant feathers. An early memory arose from that lake of light.

One day with my parent's approval, my sister Josephine received a Pinto Pony, or "Indian Horse," from a neighbor. Pinto Ponies arose from breeding domestic horses with wild horses. Though they are called ponies, they are actually horses. Pintos were the horse of choice for Native Americans.[2] Many "cowboy and Indian" movies, as well as authentic pictures of Native Americans, depict them riding these colorful, spirited horses.

Painted horses and Pinto Ponies were known and admired in antiquity for their colorful patterns. Artwork and pottery from ancient Egypt depict these horses with bold markings of white and chestnut, or of white and black and everything in between.[3] They were also common on the Russian Steppes, long before the Roman Empire's rise.[4] Pintos became particularly numerous in the United States when the Spanish conquistadors brought the breed, hardy enough to endure long ship journeys, from Spain.[5] The horses were then set free by the Spaniards before they returned home, to join with those wild horses and those kept by local indigenous Americans.

Navajo, like many Pintos, was a rich chestnut with large spots of white and a shaggy, darker-brown tail and mane. At 15.5 hands tall, he was a good-sized horse. As is common with many Pinto Ponies, he tended to be stockier and heavier than most other horse breeds. Pinto Ponies have a reputation for their strength and staying power. The most famous Pinto Pony, called "Pinto," is the only horse to have walked the United States' entire distance, a total of 20,352 miles.[6]

Pinto Ponies are known for being easygoing, easy to train, and excellent horses for first-time riders. Navajo though did not fit this mold. In fact, he was downright ornery. Josephine was Navajo's fourth owner and purchased Navajo when he was well over twenty years old. A former owner and/or trainer had apparently beaten him with a switch, and he had never gotten over it. While horse training or "breaking" methods have become more humane over the years, there is no doubt that many horses, like Navajo, have been abused through this process.

Most horses are trained between two to four years old. A commonly used method was to run the horse until it was exhausted and then start its "training." Rough handling by way of spurring and excessive jerking on a horse's mouth with the metal bit has been used by trainers, along with whipping a horse with a switch or a branch. Unfortunately, these methods were more of an expression of a trainer's social training related to controlling other living beings, and possibly of frustration at trying to achieve total dominion, than they were a suitable way to reach and teach.

Horses that were particularly "wild" or "headstrong" might be hang-tied, which is a practice whereby a horse is tied so close to a high railing so that it cannot lower its head and is kept that way overnight. This broke down a horse's resistance by wearing out its neck muscles and creating a "desired lowered head carriage." Fortunately, it is now considered an abusive practice. Navajo's ill-treatment left him distrustful of most people. Yet, if you were kind to him and he liked you, he was quite loyal. Loyalty aside, if you picked up a stick and held it up, Navajo would charge with a blood-curdling whinny, ears back and teeth bared.

My sister discovered this quite unexpectedly as she was clearing felled branches from his paddock.

I used to enjoy going down to the paddock to watch as Josephine lovingly groomed him and cleaned his hooves. With my feet perched on the bottom railing of the split-rail fence, I would rest my small arms on the top railing and admire Navajo's colorful, shiny coat. Sometimes, if allowed into the paddock, I would gently run my hand over his soft nose and through his bushy, coarse tail and mane. One day, as I watched Josephine slip a bridle over his head, I asked, "*Can I ride him?*" To which my sister, four years my senior, emphatically said, "*No. But you can come in and help me clean his stall.*" At the time, this seemed like a treat, since I could come in and pat Navajo, so I cheerfully gave a hand in raking steaming straw and manure into buckets to be dumped in a wheelbarrow outside the paddock. Josephine often said that Navajo was a "*one-woman horse.*" That he trusted her but no one else, though I think it was partially her not wanting to share Navajo with a little sister, she certainly did have an extraordinary relationship with Navajo. It must have been very healing to have an owner who loved you as much as my sister loved that horse.

Navajo and I never developed what one would call a "horse-and-rider relationship," but I knew he trusted and liked me. Perhaps he recognized that a seven-year-old, especially one who loved animals, was no threat. In any event, on some days when Josephine was not around, I would go down to the paddock and bring him treats. As his lips gathered the carrot or apple offering from my hand, I would comb his shaggy brown forelock. On some days, I would lean against his warm side and look up at the sky.

When Josephine was away at a friend's house, I visited Navajo on my own. He was in a particularly good mood and gave me a welcoming whinny. I entered the paddock and coaxed him over by the split rail fence with a carrot. He stood sideways to the fence, munching his carrot while I climbed onto the fence and swung onto his back. After sitting up for some minutes, I decided to lay down with my head and stomach resting comfortably on his back, with my arms dangling.

In the warm stillness of the afternoon sun, I even started to feel a little sleepy. Soon Navajo walked to the paddock's edge and tried to reach for the green grasses outside the fence by bending his head under the bottom rail. As Navajo tried unsuccessfully to reach the grasses, he moved close to the paddock gate that I had accidentally left open. Navajo ambled through the gate. As soon as he did so, his ears perked forward, and he lifted his head high as if hearing a distant call.

A complete change came across his calm demeanor. He shivered with excitement as he came to realize that he was free to go wherever he wanted. He then either forgot that I was on his back or, more likely, felt unencumbered by my 40-pound presence. With his head up, he walked quickly away from the paddock. Once he passed under the large oak at the corner post, he began to trot. Alarmed at the fast, bouncing pace, I took hold of his mane with my hands and clung with my legs to his bareback as best I could. It was clear that neither Navajo nor I considered myself to be Navajo's "rider." I had no control over the horse. Navajo's motivation, direction, and speed were of his choosing alone. I was just along for the ride.

Navajo trotted along the dirt path through the woods until he reached our private driveway. I shimmied up closer to his front to be able to throw my arms around his neck for a better hold. He bounded up the short incline to the drive and then started running in earnest. He sped past the gardener's cottage on the outskirts of our property and onwards to our half a mile-long driveway as fast as he could go. As Navajo hit speeds between 25-30 miles per hour, a keen focus overtook me.

Instead of fear, I became aware only of the horse and the sensation of flying. Leaning against the side of his neck, with my ear resting just above his forelegs, I heard his hooves thundering along the road and the heavy rhythmic breaths from his nostrils. The sound of hooves and horse breath filled my brain. We moved together seamlessly as if we were one. It seemed as if we were running together on a journey towards something that, once obtained, would change our lives forever. Experiencing Navajo's freedom run cemented a connection to my wild self.

Navajo was enjoying a wild freedom run. The ground disappeared as we flew unbounded by the Earth. Through this shared experience with Navajo, I escaped the self's restricting boundaries to become something greater. I experienced what could only be called an expansion of consciousness. It felt like my spirit had left the container of my body and joined, not just with the spirit of the horse, but with an eternal mystery.

My very first horse-riding experience made an indelible impression on my psyche.

As we neared the end of the half-mile-long driveway, Navajo's chest began to heave heavily. His gait shifted, and as he slowed, the movements became awkward. This caused me to slip down onto his side slightly. Yet somehow, I was still managing to hold on. I was barely clinging to him with every fiber of my being. I knew this could not last. At the end of the driveway, we entered a narrow public road. Navajo stopped as he sensed I was losing my grip. The exhilarating feeling of our wild run ended. It was then I became aware of my impossible position. The ground seemed painfully far away, and I would not be able to land on my feet from this position.

Navajo took a few steps to the center of the narrow country road and looked into the distance as if he was considering his choices: to continue to run and have me fall or to stay put. After a few minutes of contemplation on his part, a car appeared in the distance, driving slowly towards us. Perhaps the approaching car made up Navajo's mind. He turned sideways, standing in place like a stubborn statue. The driver of the car came to a dead stop about 10 feet away. He looked unbelievingly at the scene in front of him. There, blocking the narrow country road, was a Pinto Pony with a small, blond-haired girl barely clinging to his side. I do not recall how I got off the horse. Did I fall? Did the driver get out of the car and help me down?

Many years have passed since that journey with Navajo. To experience what freedom feels like deep in your bones is a lifelong gift. Over the years, I have gotten to know other horses and ridden a few, but I never became a horse-riding enthusiast in the term's typical meaning. Yet, I have always been intrigued by what lay deep within the soul of a horse, and the abiding, sometimes mysterious relationship between Equus and us.

In a cave in Chauvet, France, horses drawn over 32,000 thousand years ago come to life in vibrant colors of ochre and reds.[7] These beautiful wall drawings inside dark caves are exquisite and reflect the artistry and imagination of the prehistoric peoples who made them. Though made by firelight, the drawings depict a profound understanding of these animals and more. Besides bearing the horses' physical beauty, grace, and power, the figures evoke a mysterious, mystical quality. Some archeologists believe that some of the drawn horses were "captured" during a "shamanic trance" to gain wisdom and prophesy and that this is what gives them their otherworldly feeling.

The Chauvet cave is but one of over 250 caves that contain drawings from the "dawn of humankind." The vast majority of these Paleolithic drawings are of animals (not humans), and many are horses' representations. It is possible that horses were an important food source, and that drawing them may have been part of a ritual to bring them closer to where they lived. That does not explain why these drawings were created under such challenging circumstances deep within hidden caves. Nor does it explain their unusual mystical propensities. While there are ongoing debates about these drawings' purpose and meaning, I believe that ancient peoples dreamed a lot about horses.

In cultures that practice shamanism, the horse symbolizes the journey from ordinary to extraordinary realities.[8] Horses in many societies have figured prominently in this way. Fiery horses in Hindu and Greek mythology pull the sun god's chariot across the sky. Sleipner of ancient Nordic culture, Pegasus of the Greeks, and Muhammad's horse Alborak all manifest mystical aspects. The winged horse reflects the ability to fly to heaven. In shamanic practice, it is the horse that can bear its rider safely to the otherworld.

Imagine the first time a human being rode on a horse. That first ride must have been more thrilling than traveling to the moon. You cannot feel the wind in a rocket ship how you can feel the wind astride a horse. The dramatic leap from walking on one's own feet to "flying" 35 miles per hour exploded waves of new possibilities upon the shores of human consciousness. That ancient memory of that first ride on the back of a horse is locked safely in the human DNA.

Horses can take us on a spiritual journey. Through our interactions with them, we can reach the furthest shores of our consciousness and beyond. We can become free by joining with their spirit breaking the bonds of societally imposed limitations that hold us. In truth, we and the horse tribe are made of the stuff of stars, bathed in cosmic light from the birth of the Universe.

As my eyes drifted back to the sea of light radiating from the birds' wings below the plane, I had a premonition that there was much more that I would learn from the Horse.

II. WATER

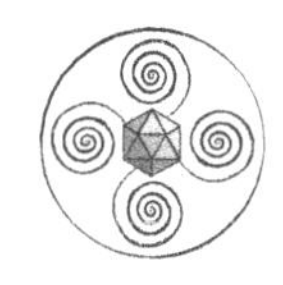

5. Selkie

In the beginning, all was water. The water is the Mother and Aluna. Where there is water, there is life. Everything can be imagined into being with the water. Water contains memory of the past and potential for the future. Without water, nothing can be imagined. Without water, the plants would die, the people would die, Mother Earth herself would die.[1]

The Kogi

Water is the lifeblood of the planet. Each body of water has its own innate essence, its own character, its own song. I recall the sparkling seashore near my childhood home and swimming in the salty waters of Long Island Sound. Across a long stretch of green lawn, past a red-bricked gazebo, and below a steep cliff lay a small, sandy white beach. Kept in place by large boulders to form jetties to stop the erosion of sand by the long shore current. Some days, I would climb down narrow wooden steps to the beach with its gentle waves lapping at the shore.

On cold days in the winter, I would climb up the boulders and practice jumping from rock to rock. Over the years, I was able to jump across some of the widest gaps without falling. On warm days in the summer, I would go for a swim in the blue waters. I spent many hours in the water, sometime swimming far around the jetty towards the light house, and other times laying on my back experiencing the delicious feeling of weightlessness with the sky above.

On many days, I would see the rounded fins of bottlenose dolphins peaking above the water line. I would hold my breath and stay perfectly still until they disappeared beneath the waves. Long after we had moved away, the dolphins disappeared from Long Island Sound. Some say it was because of pollution and increased boat traffic. Today, after a twenty-five-year absence, they have been making a comeback to the delight of residents.[2]

Water is a life-giver and life-sustainer. I have felt as much at home in the water as on the land. Surrounded by the teeming life of every kind, size, and shape beneath the surface, I felt at home. Diving beneath the water, baby flounders, half-hidden under the sand, would look up at me quizzically with their flat faces. Meanwhile, ancient-looking female horseshoe crabs would visit where the waves met the shore, with their smaller male mates clinging to their brown shell-like backs.

These arthropods come from a 445 million-year-old lineage and have been on Earth before the dinosaurs' time.[3] One of their most unusual traits is that their blood is blue due to its high copper content. Watching them, with their characteristically strong and determined movements in the fast-swirling waters, one can understand their longevity. Nearby, blue crabs, with their beautiful cerulean legs, dance with little specks of sand floating between their legs.

Nearshore, at low tide, there were mussels, periwinkles, hermit crabs, and razor clams. These clams would surprise with a squirt of water up through the sand when you walked near their holes. I collected shells and sand dollars as if they were the most valuable of all treasures. Once or twice a year, I would see the balloon-like bladders of the Portuguese man-of-war floating stealthily on the water's surface. I would swim quickly back to shore to avoid the stinging tentacles that can stretch up to 25 feet away and cause painful welts that last for days. Besides the dolphins, these were the ultra-predators able to stun and kill large fish. Now and again, a grey sand shark would get caught between the sandbars at low tide. I would dig a shallow trench in the sand to help them escape. Thus, the sea and the land were filled with beauty, drama, wonder, and mystery.

On my tenth birthday, my mother, aware of my love of the sea, gave me a little bronze statue of a mermaid. It is a replica of the life-sized sculpture of a mermaid that sits in Copenhagen Harbor. It is one of the few possessions that I still have over 50 years later. The mermaid touches a heartfelt chord in my psyche, reminding me not just of my affinity with the sea but also of a profound and poignant reality. A mermaid is a being between two worlds, not fully belonging to either one. I have felt this way at times, and the mermaid's gift was my mother's way of showing me that she understood this.

A part of me has always been "fay," as my mother would say, referring to my otherworldly perceptions and how I would "disappear" into that "place" where trees could speak and animals could tell me their secrets.

While my mother was reticent about my focus on the "wilds" and the "magical beings" outside our door, my father encouraged my interests and gave me the book Green Mansions.[4] Written in 1898 by William Henry Hudson, it is a tale about a young man called Abel, who travels into the wilds of the tropical Guyana Forest. On a walk deep into the forest, Abel hears a strange bird-like singing and follows it into an unchartered area to discover "*Rima*," a "*wild solitary girl of the woods*." Hudson's absorbing tale, of a nymph-like jungle girl called Rima, spoke to my soul. Rima was truly One with the forests, trees, and animals. She was part human part bird:

"Her figure and features were singularly delicate, but it was her colour that struck me most, which indeed made her differ from all other human beings. The colour of the skin would be almost impossible to describe, so greatly did it vary with every change of mood—and the moods were many and transient—and with the angle on which the sunlight touched it, and the degree of light.

Beneath the trees, at a distance, it had seemed a somewhat dim white or pale grey; near in the strong sunshine it was not white, but alabastrian, semi-pellucid, showing an underlying rose colour; and at any point where the rays fell direct this colour was bright and luminous, as we see in our fingers when held before a strong firelight. …

Most variable of all in colour was the hair, this being due to its extreme fineness and glossiness, and to its elasticity, which made it lie fleecy and loose on head, shoulders, and back; a cloud with a brightness on its surface made by the freer outer hairs, a fit setting and crown for a countenance of such rare changeful loveliness. In the shade, viewed closely, the general colour appeared a slate, deepening in places to purple; but even in the shade the nimbus of free flossy hairs half veiled the darker tints with a downy pallor; and at a distance of a few yards it gave the whole hair a vague, misty appearance.

In the sunlight the colour varied more, looking now dark, sometimes intensely black, now of a light uncertain hue, with a play of iridescent colour on the loose surface, as we see on the glossed plumage of some birds; and at a short distance, with the sun shining full on her head, it sometimes looked white as a noonday cloud. So changeful was it and ethereal in appearance with its cloud colours that all other human hair, even of the most beautiful golden shades, pale or red, seemed heavy and dull and dead-looking by comparison.

But more than form and colour and that enchanting variability was the look of intelligence, which at the same time seemed complementary to and one with the all-seeing, all-hearing alertness appearing in her face; the alertness one remarks in a wild creature, even when in repose and fearing nothing; but seldom in man, never perhaps in intellectual or studious man...

What inner or mind life could such a one have more than that of any wild animal existing in the same conditions?...

Why had Nature not done this before—why in all others does the brightness of the mind dim that beautiful physical brightness which the wild animals have? But enough for me that that which no man had ever looked for or hoped to find existed here; that through that unfamiliar lustre of the wild life shone the spiritualizing light of mind that made us kin."[5]

I still have the original book my father gave me with its well-worn cover and a four-leaf clover, lovingly placed between the pages.

While my father was often away at work, it was my mother that was with me the most, at home. So it was she who was "tasked" with trying to keep an eye on an inquisitive and overly adventurous daughter —no easy feat. I walked "too soon" at nine months and ran soon after. Pulled along with an absorbing interest in everything outside, I often got into precarious situations. When I was four years old, I recall looking down at her anxious upturned face after I had climbed "too high" into the top of a tall tree to put a baby Robin back in its nest.

She frequently also had to call me back to shore, when I swam "too far" out, and as a "wee girl" of eight I once disappeared into the woods for most of the day. Sometimes I was lost and sometimes I simply did not want to return home and instead preferred to be immersed in the natural world. On a family vacation to Puerto Rico, I got so excited by the new plants and animals that I saw on a visit to a tropical forest that I ran ahead and went off trail. It was hours before I realized that I had strayed far from family. Fortunately, no harm ever came to me, but my behavior must have added grey hairs to my mother's otherwise lustrous auburn locks.

My Scottish mother was strikingly beautiful. She had a lovely figure and green eyes, the color of woodland ferns. Many people thought she looked like Ingrid Bergman, and she certainly could have modeled. Yet, she was far from vain. She was a natural beauty who wore little make-up and never pierced her ears. Born in Scotland, she grew up in Edinburgh, the "*Paris of the North,*" and graduated from Scotland's oldest private girls' school, the Mary Erskine School for Girls.

She was proud of being a sophisticated "city girl," so it was no wonder that she found my exclamations about "mystical experiences" in Nature to be slightly disagreeable. Later in life, I understood that the perceptions and ideas I held were considered "backward" by those my mother's age seeking to embrace a "new modern world." It is hurtful to know that my mother's family and many of her generation and those before her denied, out of shame, their knowledge of ancient folklore and/or the mystical ways of old. Yet, considering the persecution that our ancestors faced for their beliefs, it is very understandable.

The denial of direct personal experience with what some may deem an "otherworldly existence" is anchored in a purely materialistic world view. Yet, it is a worldview that is being overtaken by new beliefs that accept that there is more to the world than what we can see, taste or touch. Even the ancient word "fey" or "fay," which generally had positive connotations in the past, was changed to a more negative turn of phrase. The original meaning, associated with fairies,[6] broadly refers to an invisible world of magic and mystery inhabited by unusual beings that only interact with certain people.

The Picts, and the ancient Celts, believed in a spiritual realm that has been around since the beginning of time. It is real yet exists in another dimension and is inhabited by beings with powers well beyond those of most humans. Since Western society has largely limited the meaning of "science" to physical and material things one can touch or calculate, there is no measuring stick for the verities of the invisible, mystical and sacred worlds. Celtic folklore and beliefs about these imperceptible beings and places are often misunderstood. In my mother's case, no doubt she and her parents, and grandparents and great grandparents were told that these beliefs were merely silly superstitions of an ignorant old fashioned people.

Fortunately, credence has been afforded to these ancient beliefs by a few, including Walter Yeeling Evans-Wentz, who decided to study and document them as best he could. In 1908 he traveled throughout Scotland to speak with the "common country folk" about their experiences and collect ancient folklore. According to Evans-Wentz, who was of Celtic descent but born in America, "*These experiences of mine lead me to believe that the natural aspects of Celtic countries, much more than those of non-Celtic countries, impress man and awaken in him some unfamiliar part of himself—call it the subconscious self, the subliminal self . . . or what you will, which gives him unusual power to know and to feel invisible, or psychical, influences.*"[7] Evans-Wentz believed that once people left country life, abandoned their ancestors' ways as backward, and lost their connection to the land, their sympathetic and responsive contact with Nature was lost, along with more subtle realities.

Yet, when I was four, my mother proudly watched, from the kitchen window, as I patiently observed a paper wasp making its nest from our wooden fence for hours. She knew that Nature was my teacher and that I understood the seen and unseen worlds. She also knew of the "danger" of contemplating and interacting with an ineffable, unseen world. The societal disparagement of such a view likely lurked in her mind, as in the minds of her predecessors. Yet, she did share with me the story she had heard as a girl, of Robert Kirk, the Scottish minister at Aberfoyle who vanished with the fairy folk in 1692.

Kirk believed that the "Sleagh Maith" or fairies were real and that they could be proven scientifically and even incorporated into the Christian faith. He researched the local people's beliefs in the supernatural world of the "otherworld" of fairies and elemental beings. He even collected several accounts of his own parishioners who claimed to have had personal interactions with them.[8] Kirk wrote: "*These Siths or Fairies they called Sleagh Maith or the Good People ... are said to be of middle nature between Man and Angel, as were Daemons thought to be of old; of intelligent fluidous Spirits, and light changeable bodies (lyke those called Astral) somewhat of the nature of a condensed cloud, and best seen in twilight. These bodies be so pliable through the sublety of Spirits that agitate them, that they can make them appear or disappear at pleasure.*"

Kirk is said to have known the faerie's secrets so well that he was carried off by them one day when he went for a solitary walk into the wilds of Doon Hill.

~

One of the few "unusual" family stories my mother felt comfortable discussing was how her father, my grandfather, Ian MacGregor, rose abruptly from the dinner table one night during the second world war and gasped that something "*terrible had happened*" to her brother Ivor. At the time, my mother was ten years old, and the German Luftwaffe was bombing England and Scotland. It was a harrowing time for Scots and all of Britain. Peterhead, a town in northern Scotland, was the second most bombed location after London. Scotland mandated sending children from major cities like Edinburgh to secret and remote countryside locations to keep the children safe. Thus, my mother was separated from her parents and sent to a farm in the countryside with several other girls for part of the war's duration, while my grandparents stayed on in Edinburgh.

My mother's only brother, Ivor, joined the Royal Air Force as a fighter pilot and flew many missions to Germany and France during its occupation. Known as "The few," so-called by Winston Churchill in his famous victory speech, "Never, in the field of human conflict, was so much owed by so many to so few." Ivor was an exceptional pilot. He once flew his plane under the Bridge over the River Forth on a dare while he was in training. A few days after my Grandfather's spontaneous outburst at the table about Ivor, they learned that a German gun had shot down his plane at the exact moment that my grandfather had stood up. Somehow, my uncle Ivor miraculously survived. He was one of the few fighter pilots to do so.

Given the horrific and frightening times that my mother lived through, it is no surprise that she had little time to contemplate the inexplicable. Over time I acquiesced to let those mystical realms fade from my life and not speak about them. Yet, when one's inclinations are blocked, there is bound to be a need for release.

As a pre-teen and young teenager, I donned the whimsical "trickster" Persona. For example, I was an excellent swimmer and could hold my breath and stay underwater for an unusually long time. This "talent" led to me winning an underwater diving competition at our local beach club and being given the nickname of "Seal." I would use this ability to shock and amaze members of a local beach club our family belonged to. The beach club had installed a small eight by eight-foot raft that members and their guests could use and swim to from the shore.

There, one could sun on its warm wooden boards or dive from it into the water. On days when newcomers or visitors were already enjoying the raft when I arrived, I would nonchalantly walk to the shore and dive stealthily into the water. I had no trouble swimming under the water, unseen, all the way to the raft. When I was just a few feet away, I would pop up out of the water with a loud inhale. My sudden, unexpected appearance would lead to startled and astonished exclamations by the raft "residents."

Other times, if I were already sunning quietly on the raft alone, when new members would swim out to the raft, and arrive and clumsily settle themselves on the bobbing boards, dripping wet and breathing heavily from the exertion, I would stand up, stretch, and dive into the water. Yet, rather than come back up to the surface, I would dive deep, turn around underwater, and come up under the raft. There, I would breathe from the air pocket between the raft and the water just inches below the newly arrived raft dwellers.

After a short time, I might hear their perplexed questioning about my whereabouts, which would rise in crescendo to eventually an outright concern about not seeing me surface. If it reached the stage where they discussed alerting the safeguard or diving in to look for me, I would swim out from under the raft and "magically appear" a few feet away. Amid gasps and astonished questions about "*where had I been?*", I would simply say that "*I was here the whole time.*"

I also used this "stealth-like" ability to "trick" people on land. I would watch people pass unknowingly under trees where I was sitting quietly in upper branches and drop a few leaves onto their heads as they passed or stood underneath. I would often say nothing but occasionally would whistle like a bird or sing to see their reaction. Rarely did anyone notice or look up, but I would act surprised if they did.

These stunts were not just fun but also motivated by a deep divide I felt towards those who did not seem to understand fully the planet we lived on. Not only were many people seemingly immune to sensing the invisible aspects of life, I strongly questioned how they could not even see the material things that mattered. For example, how could people not see and hear the alarm in the parent birds cry about its hidden nest of vulnerable babies in a tree that they were about to cut down? Or how or why could anyone fear and then club the poor, harmless sand sharks that got stuck between the sandbars at low tide? Or why did people not understand that the poison that they used to kill rodent "pests" was the reason the beautiful male red-shouldered hawk was crying piteously in the branches above, mourning the poisoning of its mate?

Everything that I loved about the Earth seemed so endangered in the face of people's ignorance or lack of care. It did not matter how much I tried to explain what I knew or saw. I was either ignored or told I was just a "silly nature girl." I felt bewildered, saddened, and angered that many people would not or could not see the world the way I did. Instead, I was told that I was the problem that I was "too sensitive" or "not living in the "real world." Yet, I could not change. I continued to rescue the orphaned and injured animals. I continued to befriend the trees. And I still prayed, almost daily, for them all.

Much of what I saw and experienced made me question the morality of those harming the animals and the trees. While it broke my heart, I learned not to say all that I felt or knew. This is true for many "sensitive" people. Those who only have eyes and ears for what they can touch sometimes denigrate the experience of those able to relate to the unseeable. Yet, those seemingly imponderable realities will, in the end, be the ones that help humankind traverse insurmountable odds and reach far beyond a limited and profane view of the world.

Sensing my struggles, when I was thirteen, my mother told me the story of the Selkie. I was enamored, even though I did not understand its shamanic roots nor its connection to our ancient Scottish ancestors, the Picts. "Selkie" is an Orcadian word for seal. Orkney, an archipelago of islands in Scotland, was once home to Scotland's indigenous peoples known now as Picts, and likely in the past as Pechts, which meant "the Elders." In Orcadian folklore, Selkies transform into beautiful humans at least once a year, usually on Midsummer's Eve. Once transformed, they are said to dance on lonely stretches of shore under the moonlight or glimpsed basking in the sun on skerries.

The Selkie-folk, as they were known locally, were gentle shapeshifters. They had the ability to transform from seals into beautiful, lithe humans. In all the Selkie myths, the shapeshifting would occur when a seal cast off its skin to assume human form. Then, instead of a seal, usually, a lithe, beautiful man or woman could be found standing on the rocks near shore. The story of the Selkie of the sea and the land spoke to me. Perhaps I too was of two worlds. I felt like a Selkie in human form who had lost her seal skin and could not return to the waters.

Over time, I learned that being in the human world and being a part of the "wild" world was not necessarily bad. Over time, I have settled into this duality, at times stepping deeply into the otherworldly realms and at others being firmly rooted in the grounded world of the day-to-day.

Much has changed in people's perception of the Earth since I first received the mermaid statue. Many people understand the importance of valuing the wilds and even believe in the empyrean forces supporting us. They have found that while we have spent the last century growing in our prowess over the material world, our understanding of how to live in harmony with Nature has not grown at the same pace. The realization that we have caused one of the greatest mass extinctions is a sobering truth that has captured thousands' attention. Thus, yesterday's "foolish childish" concerns are no longer considered childish at all but very much concerns of the "real world."

~

Memories of the waters still move me. From my earliest introduction to swimming in the Long Island Sound to sailing on dolphins' salty trails. When I was eleven, I learned to pilot a Sunfish, a small, 120-pound fiberglass boat with a single sail. These early sailing experiences taught me much about the domain where air and wind meet. It is a knowledge that has stayed with me over the years, eventually leading to my love of flying.

A Sunfish can fit two people in the cockpit, but I often preferred to sail alone. After the first lesson from a friend's father, I was hooked. I was then in the water, swimming or sailing, and had three seasons—spring, summer, and fall—to do both. Back then, in the winters, the Long Island Sound would be covered with ice several feet thick.

I spent hours learning about the wind's effect on the sail and the small boat's maneuverability. While the Sunfish lateen sail is not very inspiring to look at, at least compared to the tall, dignified sails of larger sailboats, it achieved a planing attitude at much lower speeds. Once I gained some mastery of sailing, on windy days, I would take the Sunfish far offshore to the midpoint between the Connecticut Shore and Long Island. As the little sailboat planed at a 70-degree angle, I would lean my bare back just inches over the water while sprays cascaded on my tanned skin. My hands gripped the sail rope tightly to contain and channel the mighty wind.

On one late-afternoon sailing trip, when I was far from shore, the wind began to die. The little sailboat was barely gliding on the mirrorlike surface of the water, with the sail softly flapping. It seemed like it would take hours for me to get back to shore. After trying different angles to catch as much wind as possible, I resigned myself to accepting the lull. Leaning back on my elbows, with my feet up on the side of the cockpit, I settled in. Suddenly, a huge grey bottlenose dolphin, as long as the sailboat, shot up out of the placid water like a rocket. As it did, it powerfully caught the left side of the bow, causing the boat to spin completely around like a mere toy. It is an interaction that I will recall my entire life. The sight of its 1,000-pound gleaming body and smiling face is locked in my memory.

~

There is something profoundly vital about what one can learn from the forces of Nature. It is deep learning, a teaching that reaches all the way into the bones of your body. The might of the wind, the vigor, and will of the sea, transmute your soul. It is a metamorphosis that lasts a lifetime, leaving one knowing that they owe their allegiance to primeval forces like air, water, earth, and fire. When sailing, one is guided by the sea and the wind, just as when flying a small plane, you know you owe your deliverance to the forces of the sky.

Later in life, when I had moved away to Vermont, married, and had a child, I would take my son Joseph swimming to the Connecticut River. At nine months old, he would wrap his arms around my neck, and we would swim together to the other side of the shallow, slow-moving river. There he would put out his hands to touch the smooth, warm surface of boulders that looked like sleeping whales. One spring morning, when we were the first to arrive at the river, we saw a raccoon mother with four babies clinging to her back, swimming to the far side. We similarly followed them, bonded by the flowing water.

While standing in waist-high water and gentle current, I would bob Joseph up and down. Eager to swim, he would kick his strong legs as his body rocked back and forth. He loved the waters too. When he was restless or could not sleep, I would take him to where he could hear the river's song. He would instantly relax and listen to the sound quietly and intently as if deciphering mysterious words. He fell asleep knowing that every river has its own unique song.

Parenting is not all beauty and ease. In fact, it's a challenge to keep life in balance as a mother. I wanted to spend most of my time with my young child, but we also needed to make a living to keep a roof over our heads and food on the table. Great financial and time burdens are placed on many parents, especially in America, which has some of the worst family leave practices in the world. Many parents have to work long hours away from their children. Fortunately, I was able to work at least half the time at home, and my husband and I managed a schedule so that we never had to leave our son with an outside caregiver more than 15-20 hours a week. This was a gift not available to many families; I am extremely grateful for. It also led to us making life choices leading to, at one stage, actually living for some months in a rent-free Barn.

One night, when I could not sleep, I slipped outside of our little cottage and went swimming in the large lake just steps from our door. The moon was at a half-crescent, and the sky was dark. The water was black as night and as still. I took off my nightgown and slipped into the silken arms of the lake. A low fog hung near the surface, hiding even the close lights from our cottage porch. The moon infused the fog with a subtle glow. Swimming at night is like floating in an unknowable place. The daily concerns melted away with each breaststroke as my hands reached for the water. I became enlivened by the coolness of the water on my skin, but there was something more.

Layer upon layer of care was lifted from me. It was if the water knew me intimately, allowing every unwanted or unhelpful thought and deed to be washed away without judgment. Everything that had seemed a priority earlier that day was unessential. As I let go, I felt myself getting younger and younger as years of troubles seemed to slip away in the silky cool wetness. Soon I felt so innocent and untouched by the burdens of the world that I imagined myself as a newborn. It was a sacred communion.

Later that night, I recalled the trees' songs and sayings that "spoke" of the waters. The trees would almost quiver with excitement when conveying their "knowing" of them. Of course, water is important as a singular common need among all life and its astonishing ability to take on three separate forms, solid, liquid, and gas. The trees added to this general knowledge and indicated that water is a uniquely purifying substance. Water can dissolve not only physical aspects; as we can see when we wash dishes or take a shower, it cleans "energetic and emotional layers" away that are no longer needed.

Many years later, I led a quest for First Nations women in Canada. They invited me on a canoe trip in their handmade, beautifully decorated canoe. As we women paddled two canoes to an inlet to see spawning salmon, one of the women, a leader in the community, told me that her people believed that the action of quietly dipping the wooden paddle into the dark, cool water was healing, and had the effect of washing away hurtful feelings and emotions. This is exactly what the trees told me and what I experienced myself. This knowledge sheds light on the practice of baptism, which is a water purification ceremony to wash away one's "sins."

When water is polluted, it can no longer serve those purifying roles. Many waters on the planet are now polluted. We can clean them physically, but we must also purify them energetically and spiritually by working with the invisible forces. This can be done by "praying" over the water, blessing it, or sending it good thoughts and energy. I once attended a water purification ceremony with Dr. Masuro Emoto in Los Angeles. Emoto is a Japanese author and researcher who believed that human consciousness affects the molecular structure of water.[9] He conducted experiments that showed that water responds to words and music.

When he wrote positive words on a clear glass of water, such as the words "beauty" or "love," the water formed beautiful and symmetrical crystals when frozen. Water provided words such as "hate" and "ugliness" formed chaotic and hideous-looking crystals. He found that combining the words "love" and "gratitude" formed the most beautiful and harmonious crystals. Since human beings are over 70 percent water, the messages that we receive from others and give to ourselves are important. Our group of 35 people, led by Dr. Emoto, gathered silently around a 30-foot-long water feature and "prayed" over it by sending our good thoughts and intentions directly to the water.

The water samples clearly showed that the water's structure and quality had been beneficially transmuted from polluted water, energetically in a dead-like state, to a purer and vibrant form. Our human prayers caused the crystalline structure of the water to become beautiful. Thus, the trees' shared insight about the literal "consciousness" of water, as well as some of its other qualities, is now being proven. Water can absorb, retain, and provide information that can be life-affirming, life-denying, or on a scale between.

One day, as I lay in our family's maple tree in Old Field looking up at the sky and wondering what lay beyond, the tree shared that water could be found beyond the Earth. Fortunately, I was too young to have been "taught" science. Therefore, I never questioned the tree's assertion that water existed outside of our planet, which in the 1960s was not only not considered a possible scientific reality but would have been laughed off as pure fantasy. Yet, in 2011, astronomers not only found water on other planets in our solar system, like mars, but they discovered a huge source of water near a distant quasar.[10] This gigantic water body has 140 trillion times the amount of water as that found in all of the Earth's oceans!

Beyond all of the magical and pratical abilities of water, it is an empyreal substance first and foremost. This aligns with many of the world's ancient cultures that acknowledge that water is a primordial, heavenly substance, a cosmic ocean. For example, in Hindu religious cosmology, the Universe and Milky Way are cosmic oceans that can be churned to create the nectar of immortal life known as Amrita.[11]

Trees have beautiful songs about water, timeless songs recognized by the soul. When my mother died of cancer at the too-young age of 52, I felt as if I had lost my best friend. My three sisters and I carried her ashes to the tidal river, where we all used to swim. The river flowed through an estuary and out into the salty waters of the open sound. We planned to release her ashes into the water and then say a few words from the shore in remembrance. As we stood just feet from the water's edge, each of us took a handful of her ashes and placed them into the dark, swirling water. As the ashes spiraled into the salty waters, silently, one by one, we undressed down to our underwear and entered the river.

We floated silently in the midst of my mother's ashes as the river joined the mouth of the estuary. I felt my mother's spirit joining with us and the river as we all floated together. All was understood in the river, and all was forgiven, all was purified. Before being swept out into the open waters of the sound, my sisters and I swam to the shore of a deserted sandy beach. Silent still, we climbed out and walked the short distance back to our clothes, still lying in four small piles where we left them.

Sometimes the waters can inspire one to do something unexpected, like gracefully fulfilling a sacred act with those one loves. I cannot forget the many waters, the silky feel of them upon my skin when swimming, the joy that comes from weightless freedom, or the day the river inspired my sisters and me to merge with it in remembrance of our mother. In that simple act, we became One with the sacred, my sisters, my mother, the river, and me.

In March of 2017, New Zealand declared that the Whanganui, the country's third-largest river, had the same legal rights as a "person." This is the first river in the world to be granted such rights. The designation had been sought by the Maori tribe, the Whanganui Iwi community, that lives near the river's basin, which runs for 90 miles. According to a Maori representative, "*the wellbeing of the river is directly linked to the wellbeing of the people. So it is really important that it is recognized as its own identity . . . it is an indivisible and living whole.*"[12]

His statement inspires me to think back on our ineffable water ceremony for our mother. The thought occurred to me that perhaps this "real" world was finally changing. That someday soon, the two separate worlds, which my mother and I felt were too contrary for a person to fully and wholly inhabit, would become One: a world where the sanctity of Nature is fully acknowledged and honored in the light of day.

6. Last Child in the Woods

There was a child went forth every day,
And the first object he look'd upon, that object he became,
And that object became part of him for the
day or a certain part of the day,
Or for many years or stretching cycles of years.
The early lilacs became part of this child,
And grass and white and red morning glories, and white and red clover,
And the song of the phoebe-bird,
And the Third-month lambs and the sow's pink faint litter,
AND *the mare's foal and the cow's calf.*[1]

WALT WHITMAN

When I was four, our family moved to a little town called Setauket on Long Island. Long Island is shaped like a fish. It has one "tail fin" pointing west towards the Long Island Sound and one "tail fin" pointing east towards the Atlantic Ocean. Our family lived in a large mansion on a small peninsula that jutted out like a dorsal fin into the Long Island Sound. At the very tip of the "dorsal fin" stood the Long Island Light House with its gleaming lights to warn sailors away from the rocky shore. Surrounded by sea and sky, with long stretches of woods, our six-acre property was an enchanting place for a young child to grow up. Those early experiences with the land, the water, the trees, and the animals formed me in no small way.

One of my most joyful memories was coming upon a baby squirrel sitting on the ground. When I kneeled to have a closer look at the small, furry baby with large dark eyes, he approached carefully holding his tail above his body protectively. I sat quietly for a few minutes as he sniffed me and then he hopped into my arms. I placed my hand gently on his soft gray fur and stood to walk towards home. As I walked, he climbed up onto my shoulder and sat there as if he belonged. The squirrel turned out to be a most playful companion. He loved to play tag. He would jump quickly onto my feet or legs, and then turn and run away, glancing back now and then to see if I was coming. I understood this game from watching the wild squirrels play. While I was rarely fast or quick enough to "tag" him, he learned to let me "tag" by just coming a few feet away, then he would turn and chase me. Besides the squirrel, I brought home all manner of animals and insects that needed care, including a bat, praying mantises, mice, rabbits, and many types of birds. I always seemed to find the ones that needed help, or they found me.

My parents supported my fascination with wildlife. My Welsh and English father and Scottish mother immigrated to America just six years earlier. I was the only one born on American soil, the fourth daughter. My three older sisters were born in Scotland, England, and Canada, respectively. My father was a doctor who later obtained his degree in psychiatry and opened a private practice in our home. My beautiful mother, 34 years young, stayed at home, raising us four attractive, healthy, and independent-minded daughters.

Born in 1909, my father served in the British navy as a doctor for seven years. War turned my father into someone very different from who he might have been. This is the story of most military personnel who experienced combat firsthand. He was sent on a secret mission in the 1930s to support the Spanish rebellion against Nazi-supported General Francisco Franco. After my father's death, we found an old suitcase in the attic filled with pictures he took during the war. Black and white film cannot hide the gore of death or the faces of people who have lost their faith in the goodness of humanity.

Of course, war is not just devastating to the warrior or soldier. Some veterans who were lucky enough to return home found that their children and grandchildren were born with serious congenital disabilities. One woman I know whose father was exposed to Agent Orange in Vietnam has a backward heart. Her young son suffers from the same ailment. There is probably not a single person alive today who has not been touched in some way by this great ugliness called war.

Some societies with large populations and fewer natural resources mirror cooperative cultural norms and have no wars, while other societies that may have more resources at their disposal choose to become warlike.

War is a singularly human activity.[2] Margaret Mead first indicated in her 1940 essay, "*War is only an Invention—Not a Biological Necessity*," war was a choice that some societies made.[3] For example, although Eskimos are far from meek and deal with extreme weather and competition for minimal resources, they do not war. On the other hand, Andaman Islanders, who also represent low population numbers and live with scarce resources, will openly wage war, even if their "armies" are tiny. In some cases, a society under attack from an aggressor will reluctantly pick up arms and make war to protect its survival, like the Pueblo Indians. Since war is an invention and not a biological necessity of human behavior, we choose not to come to arms. As the global human family lets go of an "us versus them" mentality to accept that we are all passengers on spaceship Earth, a world without war must become possible.

During the Spanish Civil War, my father suffered a head wound and had a metal plate surgically inserted into his skull. He experienced migraines for much of his adult life. He was treated with morphine in the military hospital long before anyone understood its addictive nature. As a physician, and later as a psychiatrist, he had full access to drugs like morphine and used these liberally to quiet his physical, and no doubt mental, anguish. By the time I was six, my father had been addicted to painkillers for much of his adult life. He was also subject to depression, a symptom common to soldiers and people with brain injuries. One Christmas Eve, in our beautiful home by the water, he tried to take his own life.

My sisters and I awoke at dawn on Christmas Day. Unable to sleep because of our excitement, we ran downstairs, looking forward to seeing our gifts under the Christmas tree. Our godmother Gillian, who was staying with us over the holidays, met us at the stairway's bottom. Her ashen and drawn face stopped us cold in our tracks. She grimly told us to go into the library, shut the door, and not to come out until we were told to.

A hush fell over us as the awful solemnity of the moment sunk in. After waiting patiently for at least an hour in my father's favorite room, the library with shelves filled with his beloved volumes of books on science, math, and classics, my older sister Anne (who was nine years older than me and fifteen at the time) opened the door a couple of inches to peek out. We all lined up beneath her to peer out the narrow crack. There was not much to see or hear at first. Then, we saw edges of a doctor's grey coat and heard hushed voices. We saw a glimpse of our mother, still in her nightgown, clasping her hands together. It was a somber and bewildering Christmas morning.

Yet, our father was to live another day. He was rushed to the hospital and had his stomach pumped. He soon regained consciousness and came home a few days later.

~

During the stressful months after my father's attempted suicide, I continued to spend the better part of my days exploring outside. Our property was bordered on the east by Long Island Sound and a white sandy beach. On the north side, a long, green grass field sloped towards the water. On the south side of the property, a wood of oaks, maples, mountain laurel, and mountain blueberries formed a border. This wood divided us from our closest neighbors one-tenth of a mile away.

Our quarter-mile-long driveway passed a gardener's cottage on the way to the small narrow road to town. When I was old enough to go to 1st grade, I would walk past the cottage, along the private drive to catch the school bus on Old Field Road. I had wonderful memories of this solitary walk along the tree-lined drive. One wintery day after a large snowfall, I made a snow tunnel that I could crawl inside and was amazed at how warm and comfortable it was.

Going towards our house, the drive split in two, forming a graceful circle leading to the mansion's entrance. Inside of the large, circular driveway was a beautiful sunken garden. Earthen walls three feet high, surrounded the garden. The "bowl" of the garden was filled with the softest grass. I used to love to take off my shoes and rub my toes and the bottoms of my feet in the pale-green natural carpet. Along the garden's edge, an arbor, heavy with grapevines, gracefully enclosed a stone path from the sunken garden to a glassed conservatory. Large maple trees stood like sentinels on the garden's edge with their broad branches reaching out in all directions. Some of the branches from the largest maple trees hung low over the sunken garden as if they were reaching to feel the grasses' softness. One of those trees, with its beautifully and gently curved branches, was my favorite.

I would climb up on the lowest branch and then go further up to the tree's midpoint about 20 feet off the ground and stand or sit. From there, I could see beyond our house and its red-tiled roof to the gray-blue waters of the Long Island Sound. I could look up at the sky and feel the wind sometimes soft and gentle, and at other times brisk and salty from the Sound. I noticed seasonal changes and admired the trees ever-refashioned leaves. In springtime, the leaves were a vivid green, filled with youth and almost glowing with an inner vigor. In the fall, those same leaves, now much larger, blazed in reds and yellows, seeming to radiate all the days of sunshine they had accumulated through the summer. The sky changed, too, with the seasons. The pale blue sky of spring turned into the deeper, darker blue of fall. I noticed how the sky mimicked how subtle hue of human eyes as they shift and change to match one's emotions.

In the late afternoons, I would lay my small body down onto one of the widest branches and listen to bird songs. Laying comfortably on the tree's smooth gray arm, I could hear the leaves singing in the wind. Occasionally I would take a nap. When dropping off to sleep, the subtle internal movements of the tree could be felt. These almost imperceptible movements hinted at the tree's active internal systems, and at times I felt like I was being rocked to sleep. I also sensed something even deeper, as if the tree had a beating heart, a rhythm of aliveness. Years later, I learned that trees have a vascular system that carries water and nutrients within it in a similar way that our own arteries and veins carry blood coursing through our bodies. Through these subtle motions and more I could tune into the tree and enter into a consciousness beyond ordinary reality. In this state, I was able to understand the tree and communicate with it.

At those moments, I felt kinesthetically connected to the energy field or aura of the tree. The tree songs spoke to me about the ever-changing sky, the dark, rich soil beneath the grasses, the joy-filled squirrels that played in its branches, and most of all about the waters: the rain that fell upon its leaves and branches, bringing good news from the heavens, and the water that lived and moved underground, nourishing the tree's roots with moisture and memories of faraway times and places. The tree was my companion and my comfort. I always felt whole in that tree; it connected me to my inner self, but most importantly, it connected me to the cosmos and to a sense of unity with all life. I remember those tree songs still.

The childhood ability to merge easily with the "biological energy fields" or "auras" of the trees, animals, and even the earth has stayed with me. It is as if my entire being is an energetic "antenna," able to pick up subtle vibrations from other living things. In deep quietude and concentration, my awareness can be expanded to be able to "feel," "see," and "hear" beyond what is considered "normal" sensory perceptions. This includes the ability to see auras around people, animals, and trees, also known as biofields. Special photography known as Kirlian photography has captured the aura, or subtle field of electromagnetic energy, that radiates from all living things.[4] In Christian art, one may see an aura around the body of a saint or a halo of light around the figure's head. Known as an aureola or aureole (from the Latin aurea, for "golden"), it is a light that emanates from the saint's body.[5] In Eastern spiritual traditions, it is known as the light from the chakras or energy centers within the body.[6] The auras of all living things often have specialized colors that depict different states or aspects of a person.

At birth, all human beings have the capacity to sense subtle energies and auras, but not everyone holds onto this way of experiencing our world. Perhaps I retained this inborn ability because I spent so much time outside with the trees and animals and had such a high degree of interest in getting to know them at the deepest level. No doubt, my respectful attitude towards different life forms helped to open the door to the possibility of learning from them. By doing so, I was naturally enhancing my consciousness's capacity to retain this more subtle ability to perceive well into adulthood. I held fast to the "wild," or natural, aspects of my being, which has enabled me to communicate and connect in this way with other life forms. I learned that they reflect many of the same characteristics that we humans ordinarily like to claim for ourselves alone.

The notion that not just animals but plants are capable of being conscious or feeling emotions was first recorded in 1848, when Dr. Gustav Theodor Fechner, a German experimental psychologist, suggested that plants have emotions and that one could promote healthy growth with talk, attention, and affection.[7] In the early 1900s, a well-respected Indian scientist, Sir Jagadish Chandra Bose, conducted experiments on plants. He discovered that plants appear to have a sensitive "nervous system" and can respond to shock by a spasm, similar to animals and people.[8] Bose found that plants grew more quickly when grown with harmonious music versus harsh sounds and claimed that plants could feel and understand emotion and intent.

Many years later, Cleve Backster, known as the father of polygraphy, proved that a special bond could be created between a plant and the person who cares for it. Once attuned to a particular person, plants appeared to stay in touch with that individual, even over long distances.[9] Backster found that when he wired a plant to a sensitive machine that measured the plant's "reactions" and compared these reactions to those of its owner, there were some amazing similarities. The plant's electromagnetic graph showed heightened emissions similar to how a person's body displays a heightened stress response.[10] For example, when he wired a plant, its caretaker, who was afraid of flying and was about to embark on a flight, he found a direct correlation between the owner's stress response taking place and simultaneous and sympathetic "stress" responses from her plant.

It was as if the plant felt its person's discomfort. In an attempt to discover how the plant communicated with its owner over great distances, Backster placed a plant in a Faraday cage inside a lead container. Neither blocked nor jammed the "communication channel" between the plant and the person.[11] This is because the invisible aspects of our world and every living thing is ever present and cannot be denied. Our science is finally catching up to this reality. Quantum entanglement proves that electrons, which are subatomic particles often bound to a nucleus of an atom, once "intimately connected" can affect each other over great distances. Human beings and plants are comprised of these electrons, thus also have this ability.[12]

Plants can do more than simply pick up on their caretakers' emotions; they can also respond to specific thoughts and intentions. For example, plants have been shown to grow more quickly when their caretakers send them loving thoughts. Plants also exhibit stunted growth when people hold thoughts of harm toward them or simply neglect them.[13] As Backster's experiments have proven, plants can and do interact with human beings at an energetic or intuitive level.

This is how I was able to understand and communicate with plants. Plants are sensitive and capable of picking up information and sensations from the world around them. Trees enjoy and are uplifted and energized by birdsong, for example. Today, it's been proven that trees planted in areas where singing birds live grow faster and more vigorously than trees planted in silent fields.[14] While there are certainly many factors that could influence this outcome, I learned from the trees that "hearing" the song of the birds strengthens a trees' life systems and encourages their growth in a similar fashion that music therapy can provide emotional and even physical healing for people.

In one scientific study, scientists at the Agricultural Biotechnology in South Korea found that rice grew faster when classical music was played in their rice fields.[15] In another study, seeds of the plant Vigna radiata were planted and then exposed to three different growing conditions: one set of seeds were left in silence, one set was subjected to a two-hour recording of discouraging words, and the third set was placed in proximity to two hours of ancient Sanskrit chants. The seeds exposed to the Sanskrit chants grew significantly larger and were healthier. While scientific studies can now prove things that seemed impossible or even "magical", humans have the ability to learn these things on their own.[16]

Not only do trees have the ability to "see and experience" the world this way in relation to other living things, but they can gain information from other trees and plants around them, and even from the air and wind. Thus, they have the ability to "sense" weather changes long before human beings. For example, plants have a high sensitivity to light and notice even very minor fluctuations in light during the day. They "know" when the summer is turning to fall and winter to spring based on the amount of light hitting the earth's surface. While humans understand this also, we cannot sense these minute changes at the same level of accuracy as plants. Trees can also sense minute moisture changes in the air and can predict rainfall. In the case of the morning glory, it opens its flower petals wide in good weather and closes and compresses them, to stop them from getting damaged from a coming rainstorm. When I was a young child, notions about plants being "conscious" in some way were considered purely fanciful. Thus, when I would come home early and say to my surprised mother that it would be a hard rain tomorrow, I knew not to mention that the trees "told" me.

Plants obtain information from other plants. Released "chemicals", volatile organic compounds (VOCs), into the air. These VOCs convey a vast amount of information. For example, plants share information such as a warning about "plant predators" such as caterpillars and animals that eat their leaves and nearby plant helpers like pollinators. Today, there is a growing amount of scientific studies confirming these capabilities. For example, over 3000 South African antelope called Kudu were found dead on game ranches in the Transvaal region. It was a mystery how they died until it was discovered that the Acacia trees they were feeding from had poisoned them with Tannin. Normally, tannin levels are at low levels in the trees and not harmful, but, during a drought, they were one of the only plants being relied upon as a food source by the Kudu. This led the over browsed trees to start producing more tannin and sending chemical messages and warnings to other trees within the region to start producing poisonous levels to save themselves.[17]

Some native trees like oaks also contain ancient memories from the time that they were a tiny seed. Some of the memories contained within them refer to the position and location of an ancestral tree. This ancestral link and information has been conveyed to me visually, to the point where I could see the surrounding habitat and climate the ancestral tree experienced. The first time I became aware of this ability sitting meditating next to a favorite oak tree. The tree transmitted not just information but a clear vision or picture in my mind of an ancestral tree and its surroundings on the side of a hill. This was one of the very first images that had been conveyed to me telepathically. Which, by the way, leads me to conclude from my experience that plants can communicate telepathically with humans, or perhaps it is the "oversoul", the cosmic container of consciousness, of the entire plant species that conveyed this information? While I do not have an answer for this, I do know that plants can learn and form memories.

Monica Gagliano, an evolutionary ecologist at the University of Western Australia in Perth, proved that the plant Mimosa Pudica memorized a learned activity for well over a month during the duration[18] of the study.[19] Rupert Sheldrake, a Cambridge-trained scientist and the author of a New Science of Life: The Hypothesis of Formative Causation, believes that there is an invisible field around all self-organizing systems such as plants, animals, humans, and even cells and molecules. This field allows information and memories to pass across space and time from one generation to the next and between individuals.[20] Sheldrake relies upon his own experiments and a Harvard study that proves that information and memory can pass from individuals to the entire species. While that research concerned rodents, my experience with the plant world indicates that a vast intelligence resides within the plant kingdom. While I have not yet come across any studies indicating a plant's "knowing" of their ancestral history, I have no doubt that this truth will be confirmed and understood in the future.[21]

Trees also know a lot about the history of the land. This knowledge extends well past their own life, as the unique seed or acorn which birthed the trees stored information from the parents. Thus, the experiences of many generations are contained within seeds. During a tree's lifetime, its education is continued as the wind, water, and other living things filled it with experiences and their knowledge. Thus, stationary trees have a way to accumulate a vast amount of information even though they remain in one place. Trees can send and receive chemical signals through the air that contains information. A tree's knowing is a different way of understanding the world from ours. It is a world that one can visit if one takes the time to "listen" to trees.

Trees are also experts at understanding the elements like water, and even light itself. For example, photons, the smallest particles of light not only contain information but act as evolutionary drivers of all life on Earth. It is said that a photon can exist for a billion years or more. This is why the light that was created during the "Big Bang" 13.7 billion years ago is still largely around. Plants, of all Earths creatures have a special mastery and relationship with light and our sun. They know how to turn light into energy through their process of photosynthesis, and they also amass information from the photons that they absorb into their being. Because of this they are able to "sense" changes and trends in our sun, and even cosmic events.

Our solar system and the planets within it are bathed in cosmic light as well as light from our local sun. Therefore, those beings that are attuned and sensitive to this light can discover the information contained within. For example, a shaman of the Dogon people of Mali in west Africa taught a French anthropologist, Marcel Giraule, that their religious symbols reflect that they understood the Sirius star system. Sirius is not only the brightest star in the sky, but the Dogon claimed that it had a brown dwarf star nearby, not yet discovered at the time of their conversation with Giraule in 1947. Yet, in 1995, gravitational studies now indicate a brown dwarf star orbiting around Sirius.[22] From my "conversations" with trees, they too have gained information not just about the Earth but about the cosmos. They interpret suns and planets as "living" beings.

One day, as I lay in one of my favorite maple trees looking up at the sky and wondering about what lay beyond, the tree shared that water could be found beyond the Earth. Fortunately, I was too young to have been "taught" that this was impossible. Therefore, I never questioned the tree's assertion that water existed outside of our planet, which in the 1960s was not only Not considered a possible scientific reality but would have been laughed off as pure fantasy. Yet, fifty years later, in 2011, astronomers not only found water on other planets like mars, but they discovered a huge source of water near a distant quasar. This gigantic water body has 140 trillion times the amount of water as that found in all of the Earth's oceans. How trees or any beings on the Earth might understand these cosmic realities is a mystery.

The trees I routinely visited seemed to know our family. Some "songs" of the tree told of my mother and father, my sisters, and me. It "saw" and understood us, not with sight but with "overall sensing." When a tree knows specific individuals, it can read even more subtle emotional and physical states. Dr. Konstantin Korotkov, a physics professor at St. Petersburg State Technical University in Russia, has developed a camera that not only photographs plant (and human) energy fields but can report on medical effectiveness remedies for specific conditions.[23] Dr. Korotkov measured responses of plants using Kirlian photography. He performed experiments to measure potted plants' reactions to the emotions and intentions of different subjects. He found significant changes in the bio-electrographic activity of the plants, depending on how they were treated. Love and kindness created an immediate positive response in the plants' energy field proving that plants can sense human emotions.

~

When I turned twelve, and my interests turned to boys, fitting in with my peers, and adjusting to my new body, Nature took a back seat in my life. My school peers' nickname of "Nature Girl" stung then. Along with all the other complications of the pre-teen years, it was at this time that the unthinkable happened. My father, a gentle and wise soul, committed suicide. I was devastated. Yet, just when faith in the very fabric and meaning of the world was most tested, I had a spiritual awakening in the woods near our home, which brought me back to the sacred essence of nature and the trees.

A few days after his death, I lay in bed, unable to sleep. Filled with despair, I left the house in the pale light of early dawn. Pulling on a wool sweater to protect from the chilling air, I went into the forest of trees by our home. Standing in the tall, silent grove of maple, oak, and birch trees, tears streaming down my face, a profound stillness filled me. At that moment, the place within that grove of trees became as vast and infinite as the Universe. Shafts of light streaming through the trees filled each thing they touched with a strange, ethereal glory until the entire space around me was lit with sparkling golden light. In the light, the trees appear luminous and oddly incorporeal. It was if they were no longer separate from the light but emanating from it. I was then enveloped in what felt like a compassionate embrace from the trees. All-time seemed to condense into a few everlasting moments that stretched for an eternity.

In that otherworldly setting of illuminated leaves, and bark and branches, I felt connected to all living beings. My hands relaxed and opened, and the ache in my heart lightened. I stood on the ground within the tender golden light and understood in an instant that death is not the end but merely a continuum. It was as if I could see the invisible threads that kept us all held together for all time within a compassionate and wise cosmos. At that moment, I knew then that all living beings—no matter how seemingly small or lost—always find their way back home. I also knew that my father's soul was immortal and that he was being held in this self-same light and, more importantly, that he was at peace.

I fell asleep amongst the trees, lying on soft, jade-colored moss, under cover of a green canopy. I slept soundly, and when I awoke, the gossamer like golden light was replaced by the sun's rays that felt warm upon my body. I lay quietly breathing beneath the trees' soft shadows, while the full light of day touched the topmost leaves reflecting the sun like flakes of quivering fire. Everything around me seemed more colorful and animated. A towhee flitted to the Earth, just inches from my resting arm, and scattered delicate leaves with nimble yellow legs. Then, in the distance, I heard the soaring call of a red-tailed hawk, and I longed to return home. I rose, gently brushed off the twigs and leaves clinging to my sweater and walked back home a different person. I was filled with an understanding of a profound truth: no matter how dark or desperate things may appear, love, peace, and wholeness always surround us. These are the invisible truths of our existence.

I was fortunate to have spent my formative years in the lap of Nature. While the unhappy childhood moments made me aware of human frailty, Nature was and continues to be my inspiration, my healer, and my truth. Nature, the animals, and the trees have revealed many things that no human being could share. I learned from them secrets about the cosmos and the Earth that I could never find in a book penned by a human hand. I sometimes wonder what kind of person I would have become if I had not had these experiences in Nature. Their "ways of knowing" were passed along to me, simply because I listened.

~

Many breakthroughs have been made in expanding our understanding of plants and their abilities, yet there is still so much to learn. Much of what I learned from my childhood companions, the trees, is a knowing beyond words. Yet, there are a few other things I can share. For example, the tree revealed that it could recognize species and individual beings by sensing "the weight of the being's consciousness." The joy-filled lightness of a chickadee's consciousness was easy to differentiate from a squirrel's playful, energetic energy. When I asked the tree about human consciousness, I was "told," that it was the "heaviness" of the consciousness of human beings that was remarkable, but that a small number of people exist who transmuted their consciousness so that it was even lighter than that of a chickadee.

Many years later, when I read about ancient Egypt, I discovered that in the ancient text, the *"Book of Coming Forth by Day,"* there was a ceremony upon one's death, called the "weighing of the heart." To the Egyptians, the heart, rather than the brain, was the most important organ as it was considered the source of human wisdom, the very center of emotions, memory, and compassion. It was believed that upon death, the "weight" of the heart could reveal the deceased person's true character and earthly deeds. The sacred object used to "weigh" the heart was a feather belonging to the Goddess Maat. Maat represented order, truth, and righteousness. The significance of using a feather to weigh one's heart shows that being lightweight and ethereal was a good thing. If the heart weighed more than a feather, it would be consumed by a monster, but if it weighed the same or less, the deceased's soul passed into a bliss-filled realm known simply as the "field of reeds."[24] This made me recall the ultimate wisdom of my childhood tree friend.

It is important to experience Nature and living things as we grow. Opening ourselves to the power of Nature helps us to understand the Great Mystery. Otherwise, we lose our sense of belonging to the Earth. We may even lose ourselves. Yet, an infinite and compassionate cosmos will never lose sight of us, and the trees will always sing their beautiful songs for our souls to hear.

7. The Picts: People of the Horse and Sea

The Picts did not 'arrive'—in a sense they had always been there, for they were the descendants of the first people to inhabit what eventually became Scotland.[1]

Collins Encyclopedia of Scotland

Forty years after my freedom ride with Navajo, I learned that my ancestors, the Picts, were no strangers to horses. The Picts were the original indigenous peoples of northern Scotland and the Orkney Archipelago.[2] They were a pastoral and sea-going people who are believed to be descended from the megalith-builders of the islands. These Stone Age people, who built the oldest structures in ancient Britain, were the architects and builders of a magnificent Neolithic temple called the Ness of Brodgar. The Ness is truly ancient and predates Stonehenge by 1,000 years. Archeologists now know that these ancient ones later moved south and west into England from the Orkney archipelago and northern Scotland.[3] As they went, they brought culture, art, and advanced agricultural and building skills with them.

The lack of weapons at Orkney sites like Skara Brae, and the Ness of Brodgar may indicate that the Picts, though proud and fierce warriors, like many ancient peoples, rarely engaged in full-on wars, but rather fought to show individual prowess, to settle individual and tribal disputes. This is in contrast to the Roman's, who sought to absolutely control defeated peoples. When the Romans and then the Vikings invaded their lands, the Picts took up arms in earnest and banded together to repel them, knowing that their entire way of life was at risk. While the Romans were never able to tame them, the Vikings were formidable enemies and may have wiped out the local Orcadian Picts and/or forced them to flee their homeland to the mainland of Scotland and join other Pictish strongholds there.[4]

The Picts' stronghold on the mainland of Scotland was longer-lasting. Meanwhile, although the Romans conquered a large part of Britain and much of the European continent, they never conquered northern Scotland's Picts. It is one reason why the Roman Emperor Hadrian built an enormous wall dividing Scotland in half, north and south. The wall, known as Hadrian's wall, was originally 15 feet tall in some places and ran for 73 miles.[5] Its purpose was to separate the Roman-held southern part of Scotland, from the areas where Picts inhabited to the north. This is the only time in Rome's history that the army had to wall off their newly conquered territory from an indigenous population. Today, at least 10 percent of Celtic people are descended from the Picts, and Pictish culture is ingrained itself in Celtic ways and lore.

One of the great mysteries about the Picts, and a significant reason for their success against the Romans, is that they were astonishingly masterful horse people. According to Julius Caesar,

"They display in battle the speed of horse, [together with] the firmness of infantry; and by daily practice and exercise attain to such expertness that they are accustomed, even on a declining and steep place, to check their horses at full speed, and manage and turn them in an instant and run along the pole, and stand on the yoke, and thence betake themselves with the greatest celerity to their chariots again."[6]

It is highly likely that their horsemanship skills arose because they were a geographically far-reaching people. Based on present archeological digs, it has been proven that the Picts and other expert horse peoples, the Scythians and Phoenicians, were closely associated. These peoples traded, blended socially, and likely shared horse knowledge and superior horse breeds.[7] Their trading "highways" were also the seas and rivers. By traveling the waterways and by moving overland by horse and carriage, they once had dominion over a huge territory.[8] If not for the horse, the

Picts may have succumbed totally to the Romans instead of hanging onto their stronghold in northern Scotland.

Horses similarly greatly assisted Native Americans in their fight for independence against the invading Europeans. Yet, unlike Native Americans, who resided on a large continent, one has to wonder why residents of an island-based society in a land of steep and rolling hills displayed expert horsemanship. It simply did not make sense, without understanding that they traveled freely from Britain to the European continent.[9] One thing is for certain, though, and that is that the Picts treasured their horses. So much so that one of the clans of the Picts called themselves the "Epidii," meaning "the people of the horse."[10]

Horses are truly one of the most admired and revered species for their speed, strength, and beauty. These animals have propelled forward the evolution of *Homo sapiens*. The horse shaped our ancestors, physically, emotionally, and spiritually for thousands of years. The horse provided transportation, muscle power, and distinctive military advantages. Horses made distant journeys possible. We still refer to horsepower when considering a combustible engine in a car or plane, like the single-engine plane I flew above the Laguna Madre. This reflects our undying subconscious connection between travel, new horizons and the mighty horse

Besides the horse, the Picts also had other advantages over the Romans. The Picts' cultural norms and even fighting strategies were very unusual to the Romans. Romans fought in disciplined, large groupings. The Picts might attack in small bands, then retreat to safety, and then return again when the Romans were less ready.[11] This type of guerrilla warfare was very effective against the regimented Roman army. We now know too, from old swords and fighting paraphernalia, that many of the Picts were left-handed and held their swords in their left hands and their shields in their right. Thus, in face-to-face combat, the Romans would have been blindsided, expecting swords and blows to come from the right when they were attacked instead from the left.

Pictish communities tended to be small, cohesive groups that built structures, domesticated animals, and had gardens with plenty of room between villages.[12] They were adept at living off the land, preferring open country, and a pastoral tradition that they had carried on for thousands of years. Thus, the Romans, who tended to congregate in large cities, could not focus on one or a few places to conquer. Roman soldiers were further befuddled by the fact that Pictish women could be warriors. Thus, there were fewer instances of Romans coming across unarmed, helpless women.[13] Women and young girls learned how to hold a shield to protect themselves at a young age, and how to use a spear and sword against their enemies. This was astounding to the Romans, who treated Roman women as second-class citizens and largely as the "property" of their fathers, husbands, and/or brothers.[14]

Cicero, a famous Roman philosopher, wrote that "*all women, because of their innate weakness, should be under the control of Guardians.*" In Rome, women "*needed to be protected.*" so coming across armed and battle-ready Pictish women must have been a cultural shock. Rome's disdain for female warriors followed their societal beliefs about women's status and their place in culture. This particular schism of cultural perception about the "proper" place of women in society laid a singular path of destiny for the great warrior queen Boudicca.

Boudicca was a Welsh queen of the Iceni people.[15] The Iceni were living in ancient Britain when the Romans came to conquer those lands. We know the Iceni were related to the Picts because their Brythonic language, derived from Pictish, is a pre-Indo-European language from at least the Bronze Age. (The name "Brythonic" was derived from the Welsh word Brython, meaning an indigenous Briton as opposed to an Anglo-Saxon or Gael.)[16] In an attempt at forced diplomacy, Boudicca's husband Prasutagus included Emperor Nero of Rome in his estate.[17]

In return, the Romans deemed the Iceni their allies and promised not to overtake their lands and people. When Prasutagus died, he left his estate instead equally to his wife and daughters following Pictish ways. Disgusted upon hearing this, Nero's procurator ordered the Roman army to seize all of these lands. Roman soldiers plundered the estate as a prize of war and told the Iceni people that they were not allies but mere Rome captives.[18] Boudicca, now considered the Iceni's leader, objected vocally, and in return, she was publicly flogged and humiliated while her two young daughters were raped. Her punishment no doubt reflected the disdain that Roman men had for women being treated as equals.

While the Roman soldiers thought they left Boudica and the Iceni as "docile and submissive" people, they were sorely wrong. Boudicca mounted a revolt against the Romans with the full support of the Iceni people. After gathering a large faithful army, she successfully defeated the Roman Ninth Legion and conquered modern-day Colchester, and even Londinium (modern-day London). It is believed that between 70-80 thousand people were killed in these horrific battles, as these two towns were destroyed. While the Roman army eventually defeated Boudicca and her army in 60 C.E., Catus Decianus, the procurator of Roman Britain at the time, was blamed for unnecessarily enraging the Iceni people and prompting Boudicca's vengeful actions. Decianus escaped to Gaul and was soon replaced and disgraced.

With Ancient Britain's defeat at the Romans' hands, not just land ownership changed; so did cultural norms. The more egalitarian status between men and women shifted abruptly. Roman women could not vote or hold political office—so it would be for the Pictish and Welsh women under Roman law.[19]

~

Beautiful and elaborate horse jewelry thousands of years old has been unearthed in Orkney. While most people of that time reserved their ornate jewelry for high-ranking people, Pictish artisans created ornate and beautifully fashioned riding gear for their horses. Picts bedecked their horses in this way to show their ultimate respect for the horse.

Today, on South Ronaldsay, an island in the Orkney Archipelago, the residents hold an annual celebration known as the Horse Festival.[20] This festival can be dated back two hundred years, but many people believe that its roots go further back in time. During the festival, young girls dress up like horses. They wear beautifully adorned mimic horse harnesses, heavy black shoes as "hooves," as well as fringes and fur to represent the feathers of a heavy horse. Some of their horse costumes are over 100 years old and are handed down through the generations year after year.

Young boys also have their part to play and compete in plowing matches. Boys as young as five mimic horses and "plow a field," creating dreels and furrows with a miniature single-blade plow. This tiny plow may have also been handed down for many years, with many generations of young boys designing their own plow pattern. While the festival highlights the horses' domestic uses, the great Orcadian poet Edwin Muir saw the greater promise in the Horse and how it links humanity to its inner wholeness after a catastrophic war in his poem the "The Horses." His poem evinces the sense of a sacred bond between the horse and mankind.

Late in the summer, the strange horses came.
We heard a distant tapping on the road,
A deepening drumming; it stopped, went on again
And at the corner changed to hollow thunder.
We saw the heads
Like a wild wave charging and were afraid.
We had sold our horses in our fathers' time
To buy new tractors. Now they were strange to us
As fabulous steeds set on an ancient shield.
Or illustrations in a book of knights.
We did not dare go near them. Yet they waited,
Stubborn and shy, as if they had been sent
By an old command to find our whereabouts
And that long-lost archaic companionship.
In the first moment we had never a thought
That they were creatures to be owned and used.
Among them were some half a dozen colts
Dropped in some wilderness of the broken world,
Yet new as if they had come from their own Eden.
Since then they have pulled our plows and borne our loads
But that free servitude still can pierce our hearts.
Our life is changed; their coming our beginning.[21]

The role of the horse in the betterment of the human race cannot be overstated. The horse has aligned its fate with ours. The horse propelled humans in their outer and inner journeys. A site once reached through a day's, weeks' or a month's journey on foot would now be reachable in hours or days. It was the horse that revealed to our ancestors' eyes never-before-seen vistas. The horse enhances not just our physical prowess, but also our inner conceptual landscape, broadening it to reveal new horizons at the edges of our evolving minds.

The horse can be an ally to humankind's journey into an unknowable future by helping to free our minds from modern societies' numbing over-domestication, and work subject to the tyranny of industrial machines. Ultimately, through the window of a horses' soul, we may find that we have always been truly free.

8. Women and Nature

She rules the elements, the air, the earth, and the sea.
She governs the life of the animals; she tames the wild
beasts and prevents their extinction . . .
She assists in birth.[1]

The Ancient Greeks on the goddess Artemis

Before I was old enough to fly, I pretended to fly standing on a cliff above the Long Island Sound. With my blond hair tossing and my arms outstretched wide into the wind, the roar of the waves below echoed a thundering roar, leaving a white frothy foam in their wake. Where the tumultuous water meets the stalwart sky, I imagined flying, suspended between them. It seemed as if anything was possible in those early days. However, as I grew from girlhood to womanhood subtle hints of limitations began to take hold.

Seven years later, as I sat at a dinner table with my parents, "devout" atheists, I realized that little did they know of the powerful influence of religion on my young girl's heart. While my parents believed that traditional religions are wrongly used to justify prejudice, hate, and wars, the surrounding societal norms said otherwise. When I was 14 years old, I asked my teacher why there was a male God but no female God. I told her that I found this odd since women populated half of the earth. She appeared astonished that I would ask such a question and then said,

"Well, that's just not the way the world is. Remember, too, that it was Eve and her weakness, which led to the downfall of humankind." I later made fun of her answer to a classmate, but in truth, her response was not so easily dismissed.

Here was a woman, an intelligent teacher, who was comfortable relegating women to second-class citizenship in a heavenly hierarchy. Her response may be supported in the Bible, but not her by profession. In reading the bible, she must have come across passages such as "*A woman should learn in quietness and full submission. I do not permit a woman to teach or to have authority over a man; she must be silent, for Adam was formed first, then Eve. And Adam was not the one deceived; it was the woman who was deceived and became a sinner.*"[2]

The absence of a female God influenced the teacher's identity and even my own. Though I was raised "without religion," and we did not attend church except for Christmas Mass, our society's dominant religious dogma influenced me. While many Christian dictates are positive and welcome, especially when they directly refer to the words of Jesus Christ—*"Do Unto Others," "Love Thy Neighbor as Thyself"*—others are extremely harmful. For example, the existence of a male God, absent a female counterpart, sends a strong message. If divinity is deemed to flow from this one masculine source, then all beings that do not resemble the masculine God, including women, children, and animals, are demoted in this "heavenly hierarchy."

I began to wonder how religious dictates about women being the non-sacred sex might affect how women and girls perceive themselves. Certainly, religious doctrines that deny the divine feminine support the assumption that women are less valuable than men. It is easy to see how this interpretation is integrated within many societies by looking around the world. Inequality between the sexes is inherent in many world cultures. In Saudi Arabia, for example, women are legally submissive to their spouses, brothers, or uncles and must defer to them for many basic rights and privileges.[3] They must receive their male guardian's approval to leave the country or get married or divorced. The country has since 1979 enforced a strict interpretation of Islamic law called Wahhabism. The World Economic Forum declared that Saudi Arabia is one of the most unequal countries for women worldwide.[4]

Globally, women are still paid less than men, averaging 60 to 76 percent less pay than their male counterparts.[5] Women hold less than 20 percent of the land globally, yet many women work the land and tend the fields. A[6] hundred million women from over five countries are married under the age of 18, and of those, 250 million were married under the age of 15 years—some were married as young as 5![7] One out of three women on the planet will experience some type of violence in her lifetime.[8] Even in a comparatively "democratic society" like the United States, women were not given the right to vote until 133 years after the Constitution's signing and the passage of the 19th amendment. Only the last three to four generations of American women have had this right.

Yet, women fulfill so many important roles in society. For example, they produce more than half the food grown globally.[9] Women are still considered the primary caregivers, at home, and in careers, spending twice to ten times more time than their male counterparts to care for the elderly, the sick, and children.[10] They do so often without economic benefit or at meager wages. Despite the value they bring to society, women simply do not have the same political, economic, and reproductive rights and freedoms as men.

A significant pattern emerged: in societies where women were treated equally, society seemed healthier and provided a better lifestyle for its citizens. There were smaller economic divisions between people and more care for the elderly, children, and infirm. In contrast, in societies where women were not treated as equals or neglected or abused, a severe national rule was likely along with frequent wars, slavery, and great economic disparity. It also became clear that more patriarchal and warlike societies purposefully destroyed civilizations that honored feminine forms of divinity.

The Romans, for example, had an extremely patriarchal society. Women were not allowed to hold property in their name and were largely "owned" by their father, husband, or brother, needing to seek their approval on all major life decisions.[11] Rome was a conquest-hungry society that wiped out many peoples or sought to change their culture and beliefs. Measures were taken to ensure that societies that treated women as equals would be stopped, not just for a lifetime but for all time to come. This was the case with the many thousands of people who worshipped the feminine goddess Artemis.[12]

How great was Artemis before Rome and the creation of the Roman Catholic Church? One of the "Seven Wonders of the World" was built in her honor. A magnificent temple, four times the Parthenon's size, with 127 enormous pillars 60 feet high, was constructed for the goddess in 650 B.C.E. at Ephesus (modern-day Turkey).[13] The temple was surrounded by beautiful grounds, groves of trees, and a peaceful lake. Priestesses managed the temple, passed on important rituals and spiritual knowledge, and performed healing services for the community and for pilgrims. The temple grounds were also a sanctuary for those fleeing persecution.

Not only was the Temple of Artemis one of the "Seven Wonders of the World," but it was, in fact, considered the greatest of them all. Antipater of Sidon, who visited all of the seven wonders, had this to say about the monument:

"I have set eyes on the wall of lofty Babylon on which is a road for chariots, and the statue of Zeus by the Alpheus, and the hanging gardens, and the colossus of the Sun, and the huge labor of the high pyramids, and the vast tomb of Mausolus; but when I saw the house of Artemis that mounted the clouds, those other marvels lost their brilliance, and I said, 'Lo, apart from Olympus, the Sun never looked on aught so grand.'"[14]

~

Hundreds of years after Antipater of Sidon's visit, Roman Catholic priests viewed Artemis to be their greatest rival. The Bible tells of how the apostle Paul goes to Ephesus specifically to admonish against Artemis's worship. He "*was not timid about using tactics in the war to topple and dethrone the cult of Artemis.*"[15] One of Paul's complaints against the goddess was that she was not a "proper" God, born of a human body. The fact that Jesus was also born of a human body did not seem to matter. Paul was also incensed by merchants selling likenesses of Artemis to pilgrims. This is ironic given that the sale of religious artifacts, jewelry, and statuary related to the Catholic Church is to this day "big business," pulling in millions of dollars a year!

Artemis' many followers would have considered their ancient sacred practices and traditions—which had survived thousands of years longer than Christianity—worthy religious and spiritual beliefs. Yet their faith was then, and still is, called a "cult." Catholic Rome's intent to extinguish the very idea of feminine divinity for generations to come was successful for many hundreds of years. This was accomplished not just by war and conquest but also by blatant and subtle propaganda, teachings and even language.

Take, for example, the word "trivia," which refers today to "*unimportant matters*," or, as the thesaurus states, "*a collection of insignificant or obscure items, details, or information*."[16] The original and historically accurate Latin meaning of tri is "three," and that of via is "path."[17] "Three paths" to me indicated something of importance, so I decided to research the word further. What I found was quite revealing. In ancient times, "trivia" referred to the feminine sacred trinity. The Threefold Goddess is found in different religious contexts and has been known by various names in time, including but not limited to "Hecate-Selene-Artemis."[18] The multiple identities revealed different aspects of the One Goddess, in much the same way that the Father, Son, and Holy Ghost mirror the threefold identities for the Christian "God."

These three aspects of the Goddess reflected her dominion over the three realms: Earth, the heavens, and the underworld. The Goddess's threefold aspect has been identified most closely with Artemis, once referred to as "Artemis-Trivia." "Artemis" is a pre-Greek name from at least the time of Ephesus. By Roman times, Artemis was renamed Diana and had lost much of her three-fold meaning and significance.

To non-Christians, or "Pagans" like my ancient ancestors,[19] divinity came equally in feminine and masculine forms. From what we can tell today from folktales and ancient writings and sites, the Picts practiced goddess-worship and were devoted to Nature. The goddess lived among the people, and the land was venerated as one would venerate the home of a deity. Some sites on the land were considered especially divine. Supernatural sacred powers emanated from places like groves, springs, and special rocks. These were places that the goddess was especially fond of and where she conducted miracles.

This view of the feminine as a sacred vessel reflected women's role in Pictish society. They could own property and pass that property down to their offspring.[20] Women had freedom in marriage and childbearing. The Picts' custom of trial marriage provided that a man and woman could be married for one year. If either wanted out of the marriage before the end of the year, they could do so without recrimination of any kind. Even the important access to succession in kingship could be granted via the mother and matrilineal lineage.[21] Thus, there were no "bastards" in Pictish society because all children knew their mothers and could inherit from them.

While much evidence of the worship of the goddess Artemis by the vast majority of pre-Christian people has been destroyed, the surviving depiction of the beautiful statue of Artemis of Ephesus hints at her great beatitude. The Artemis of Ephesus stands warmly welcoming, with her arms held open. She bears many rounded breast-like shapes, which accentuate her nurturing aspects and her great generosity. She is clothed depicting her virginity, and she wears a collar of the zodiac representing her wisdom of the stars (the gods) and their heavenly dance. She is an inspiring symbol of abundance and life. Animals, wild and domestic, surround her body, and bees, honored for their honey and its practical and spiritual uses, are welcomed there as well. This is a life-generating image of a God.

The Queen of Heaven was clearly much beloved by many. But a tidal wave of religious dogma, initiated from Rome, forcibly swept away their faith and eventually sought to claim the very memory of a beloved feminine god. Along the way, the meaning of the word "trivia," referring to the three aspects of this powerful feminine divinity, was altered to such a degree that she would not only not be missed but forgotten as something, well, trivial.

Yet here she was, hundreds of years later, looming large in my life. The more I read and learned, the more I considered new possibilities for myself and women in general. I changed the password on my work computer to "Artemis." While the name change took but a moment, it reflected how powerful psychic shifts were taking place that brought me back to remember the core of my identity.

As the goddess of the wilds, she was known as Artemis, "Agrotera," or "untamed."[22] Her constant companions were the deer, the hound, and the horse.[23] She was considered the mother to the bears, a relation to wolf and lion, and the keeper of the willow, walnut, and cedar trees. As the Great Mother Goddess, she was linked to the honeybee and even to light itself in her earlier incantations. To the Greeks, Artemis was the goddess of wild places. Her domain was far from cement pavements, streetlights, and the noisy intersections of cities. She thrived among woods, streams, wild animals, and plants. These were the very same places where I felt most at home.

There, in the midst of the city, golden visions from a timeless realm reached me: the voices of trees and the secrets that the animals revealed were heard once again. Memories of childhood mystical experiences returned. Magical moments of communicating with the animals and trees returned. I longed to feel the grass beneath bare feet, and I picked up those sacred threads to weave a new life and create new work that reflected an inner calling. A calling for humanity to respect Nature and balance the universal forces on the Earth we know as masculine and feminine.

At the heart of the problem is that not just women but also the animals and the Earth are not considered sacred and are instead defined by their use in increasing GDP. For example, in ancient times, masculine role models were very different than they are today. Cernunnos, the ancient male deity of the Celts, was largely respected for his regenerative powers. With his antlers, Cernunnos is a blending of the wild aspects of his masculine nature and his human-like form. He embodies a deep communion with the wild wisdom of the world. As such, he was seen as a bringer of fertility, abundance, and friend to the forest's animals. He thus would make a perfect consort to his feminine counterpart and equal, Artemis.

A society that places profit over life harms all its citizens, not just women. No matter how many laws one writes, if a society does not inherently value men, women, and children, they will be neglected (or worse). When humanity's highest functions are hidden to turn the vast majority into consuming cogs in the wheels of commerce, we not only forget our higher selves, we trample our sacred Earth Mother.

In the 1970s, the eco-feminist movement first linked the exploitation and degradation of the natural world with women's subordination and oppression. The connection between the feminine and the Earth is no accident. It is the Earth that supports all life, providing nourishment and shelter, just as women and mothers have a direct vested interest and role in reproduction and birth, ensuring the continuation of our kind. Perhaps another reason for this is that three generations of women reside in one body for a period of months. When a woman is pregnant with a female baby, that baby develops all the eggs she will ever have in her lifetime before she is born.[24] Thus, there is a true physical link and continuity between the three generations of women: grandmothers, mothers, and granddaughters.[25]

Today, we no longer live in a world of the Goddess, and very few people are aware that the "greatest temple on Earth" once honored a female god. Instead, we are faced with a world where not just Artemis is forgotten, but the Nature that she loves is sacrificed to greed. Yet by valuing women and "feminine" traits of compassion, nurturing, and caring for future generations, we can begin to heal what is broken.[26] This does not mean that we lower the value of the masculine, but rather that we bring equal value to the feminine in all that we think and do. Understanding that the feminine is on equal grounds with the masculine helps women contemplate the divine power within themselves, and men to honor the feminine qualities of intuition, regeneration, and as protectors of future generations.

Societies that respect the sacred feminine tend to support women and offer greater freedom for men and children. These societies represent a balanced and equanimous approach to life, honoring both masculine and feminine characteristics, the yin and yang forces of the Universe. Today, women are gaining a larger conception of themselves as creatives, visionaries, and leaders. No longer considered second-class citizens, women can step into authentic society roles while following their paths with heart. This bodes well, not just for women, but for all people and the planet.

III. FIRE

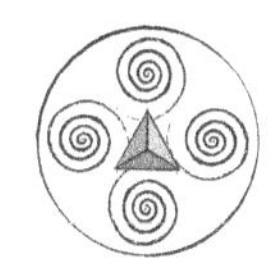

9. The Real People

Whilst the fierce wolf from which they fled amazed
'Leaves his stamp visibly upon the shore
Until the second bursts;–so on my sight
Burst a new Vision never seen before.[1]

PERCY BLYSSE SHELLEY

As the second tear slipped down my wet cheek, a silvery shadow moved behind the trees outside my hotel window. The evening light made the pine trees outside seem taller as they cast shadows on the white snow. As the tears froze in place, I found myself staring into the searing yellow eyes of a large gray wolf. It stood like a noble statue, no more than ten feet beyond the large bay window. I was in Alaska, home to the largest remaining population of gray wolves in the United States. The wolf's eyes were gold and shone with an ancient knowledge gained from its species' 30 million years of living on Earth. Its gaze penetrated my soul and pulled

the sadness from my heart, leaving a serene emptiness. I heard the words "*With a burning fire, I will make you whole.*" Feeling drowsy, I lay down onto the bed just behind me and fell into a heavy sleep.

When I awoke the next morning, I was curled up on top of the bed covers, fully dressed and chilled to the bone. Realizing the time, I jumped up and quickly changed into fresh clothes. I went to the bathroom to brush my hair, and as I looked into the bathroom mirror, I saw that my face looked lightly burnt as if I had a sunburn. It was then that the bottomless, searing eyes of the wolf leaped back into my memory like a purifying ray of white-hot light. I shivered and walked away from the mirror, saying out loud, "*It was just a dream.*" as if saying so would make it true. I stood again at the window and stared beseechingly at the dark green trees outside as if they would concur. But I could not ignore the marks in the snow leading from the forest towards my room.

Since I often see animal tracks while hiking, I knew unmistakably that these "maple leaf" shaped marks belonged to an enormous wolf, Canis lupus. I shivered again, turned, and walked over to the sink. I took the clear glass cup from its holder, filled it to the top, and quickly drank three glasses of water. I had often experienced physical sensations, sights, and sounds when connecting energetically and psychically to animals and trees. Sometimes I could see multicolored auras around them or hear what I refer to as their frequency or "song," or even feel their presence energetically, but this was a completely new sensation. However, I did not have much time to ponder my experience.

It was eight in the morning, and the International Bering Sea Conference would be starting in an hour. After missing dinner, I just had time to get a quick breakfast and get my notes for my opening speech. After leaving the environmental law career behind, I was recruited to serve as the executive director of an international environmental organization protecting species and habitat in Alaska, China, Russia, and Indonesia. Through investing in and supporting local communities, we helped hundreds of environmental NGOs, indigenous tribes, and scientists protect their environment and many animal species. We also hosted conferences and workshops where leaders could discuss their work and findings, share plans, adopt resolutions, and form powerful coalitions.

Two years later, I launched the International Bering Sea Forum Conference to bring together an international coalition to band together to protect the Bering Sea. The Bering Sea is a deep sub-arctic sea at the Pacific Ocean's northern tip, bordered by the Aleutian Islands. Every spring, the Bering is the destination for polar bears, sea otters, and marine mammals like stellar sea lions, walruses, bowhead and beluga whales, ribbon, and ringed seals, along with thousands of birds like crested auklets. The Bering supports over 12 million seabirds, providing them with a safe place to make their nests and feed their young. Its rich phytoplankton blooms are a major feeding destination for animals in winter, spring, and summer. This is why the Bering is considered the "mother-load" of marine biodiversity.

Yet, as global warming heats the Bering's normally cold, nutrient-rich waters, this, along with overfishing, is causing the Bering to lose its diversity of life. It is no longer able to support the numerous species of animals, birds, and people who have relied on it for thousands of years. Many Bering species are disappearing or are gone. In their place, noxious invaders like poisonous algae and jellyfish are making the Bering their new home. Alarmed about overfishing, poaching, pollution, and ecosystem shifts in the Bering Sea, an international coalition of scientists, managers, and community leaders has formed to push the United States and Russia to work together managing what may be the most productive marine area on Earth.

At one time, an arc of islands that form the southern boundary of the Bering served as a land bridge for humans and animals to travel freely from one continent (now Russia) to the next (now the United States). Today, the sea is divided by unnatural political boundaries. Due in part to the Bering Sea's co-owned status, countries rush to grab whatever resources it has to offer first. Scientists, indigenous leaders, NGOs, and policymakers united as an independent voice for conserving the Bering. Before then, governmental agencies and corporations more interested in the economic benefits held sway over decision-making. Out of concern for the impending demise of this once pristine and vibrant habitat, our international community worked together with the priority to preserve and protect it. A statewide media outlet did a story about our efforts:

The International Bering Sea Forum will marry scientific data with traditional Native knowledge while gathering reports from regular people who wrest a living from the ocean on both sides of the border. The *Bering Sea* fish and mammals *"don't understand political boundaries,"* said Catriona [MacGregor], speaking during a telephone press conference from the group's offices. *"They move freely from one side of the Bering to the other, so any one country's management or lack of management affects all species; this is why we need to work together."*[2]

Understanding the important historical, environmental, and cultural knowledge held by native peoples, I invited many indigenous leaders from both sides of the Bering—from the United States and Russia to participate. At least one-third of conference participants represented native communities, with the other two-thirds being represented by scientists, policymakers, and environmental leaders from five different countries. These diverse participants arrived with disparate perspectives and seeking independent objectives.

To the international policymakers, the Bering was to be regulated so that the marine "commodities" could be divvied up between countries to their benefit. To the scientists, the Bering was to be measured and studied, especially now that biodiversity was rapidly changing. To the environmentalists, the sea was to be protected from overfishing and polluting industries. Bans on large-scale fishing, ocean dumping, and strategies to end global warming were their goals. I believed that the only way the Bering could be properly protected and cared for was by uniting an empowered community of diverse people who "knew" the sea and invested in its welfare.

I was sitting near the back of the room on the second day of the Bering Sea conference, the same day as my encounter with the wolf. At the front of the room stood a policymaker, and 5th presenter, from Washington D.C. discussing the Bering's "resources." As the speaker droned on about national commodity rights, a tall, stately woman stood abruptly from her chair. Her long glossy black hair fell gracefully around her back and shoulders and spread shimmering onto a jacket of beautiful geometric patterns in red and green. With her body slightly shaking from emotion and her voice breaking, she said, *"This is our mother you are speaking of, not resources for your use."* The presenter lowered his pointer and peered dimly over his glasses in her direction, his mind struggling to make sense of what he had just heard.

The woman was indigenous to Alaska and of the Yup'ik. Yup'ik means "real people." Yup'ik culture depends upon cooperation and working within natural laws that rule the delicate Alaskan bioregion in which they live. Indigenous peoples who attended the conference, from both sides of the Bering, Russia, and the US, were losing their ways of life that had begun thousands of years ago. To the native peoples from Russia and Alaska, the sea is a living entity that supports them. She is a place not just of caloric nourishment but also of cultural guidance and spiritual inspiration.

While there had been many well-prepared presentations that day, mainly by professionals steeped in policy, economics, and science, it was the woman's simple statement that resonated in the hearts of all who attended. The room fell silent as each person felt the truth of her words settle upon them. We all knew and felt the words of *a "real"* person. Her statement conveyed the primal sacred relationship between humans and the natural world.

Many indigenous peoples are palpably connected to the planet. They know that they belong in a larger natural order. The change of the seasons, a stars' movement, a hawk's call, and the howl of a coyote mean something greater to them than to industrialized people. By reading and understanding the subtle signs of Nature, they are empowered to know the meaning of natural cycles. They live in a world with awareness and knowledge of many life forms, all connected in a dance of creation. One Yup'ik young man shared with us that he always knew when the salmon were coming home to local streams when the yellow flower, the same yellow color of the salmon's eyes, bloomed. Today, the flowers were blooming earlier due to unnatural warming trends, missing a step in the dance. He was close to tears as he said: " *My father had fished for salmon in the region for thousands of years and I cannot recall a time when my ancestors cannot not fish. But now there are so few, my family and I cannot survive.*"

The Yu'Pik woman's words rang into my being and, no doubt laid an invitation for the spirit of the Wolf to visit. Over the years, of working 60 hours a week and more and spending less time outside with the nature I loved, I had lost my own connection to the wilds. Even I, "Nature Girl," was not free from a societal bias that alienates people from Nature. Although I was working to protect the environment, the day-to-day world of spreadsheets, budgets, mortgages, car payments, and being on the computer much of the day, my relationship with Nature had waned. The trees and animals were not as much in my thoughts or mind, and even I started referring to the Bering and her "resources." I was thinking about nature and not feeling her. Yet, after Wolf's visit, I knew that as long as we objectified the Bering Sea with our societal lens, we would never see it, never know it, never really touch it or learn its secrets. It is a teaching that Wolf burned into my consciousness. Wolf, the great purifier, banished the profane thoughts from my soul, and it made me cry.

We have profound relationships with Nature, the earth, and the sea. The salmon are not just "resources" or "commodities" for our use, just as human beings are not really "consumers." We are much more than that, and so is the Earth. Sometimes we need to be shocked out of societal imposed ways of perceiving that are limiting and even destructive to nature and us. As we turn away from objectifying that which we belong to and embrace it with compassion, we not only can change the world for the better, but we will find that even we can become the "real people."

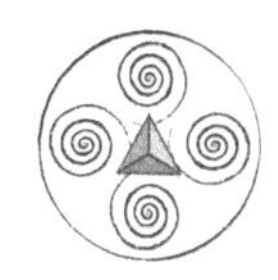

10. The Rise and Fall of "Consumers"

Earth provides enough for every man's need but not for every man's greed.[1]

MAHATMA GANDHI

Everything we make or use starts as a gift from the Earth, including gold, silver, water, trees, air, and soil. Yet, the true value of these is ignored. There are no safeguards on protecting them because clean air, water, trees, and pollinators like bees do not count in our modern economic policies and equations. One cannot eat money or build a home out of it, yet we treasure money more than the real things we need to survive, like clean air, clean water, trees, and fertile soil. According to E. F. Schumacher in Small is Beautiful, "*A businessman would not consider a firm to have achieved viability if he saw that it was rapidly consuming its capital. How then could we overlook this vital fact when it comes to the Earth?*"[2] If countries counted "natural capital" and even its destruction from industries like oil, coal, and natural gas, they would operate at too much of a loss to be economically viable. They simply would not exist!

At the International Bering Sea Forum opening, a local priest married to a Yup'ik woman told a story important to the Yup'ik people and many other native peoples. In the story, known as "*Through the Eye of the Needle*," a Yup'ik youth called Amik goes on his first hunting expedition alone.[3] As Amik leaves his grandmother's humble home and heads towards the sea, he knows his solo journey is part of becoming a man. During his travels, he comes across many animals that he kills. Amik considers bringing back what he has gathered or hunted to share with his Grandmother and his people, but instead, he eats everything himself. In an odd twist, as Amik eats and consumes, he can eat more and more.

A large part of the story's telling is a detailed description of all of the animals that Amik hunts and eats. Thus, to tell it well requires a good hour or more. Jumping ahead here, in the end, Amik is even able to eat an enormous whale all by himself! Near the end of the day, after consuming enormous numbers of animals, Amik realizes that he has turned into a towering giant. Amik becomes homesick and heads back to his grandmother's house. When Amik arrives, towering over the treetops, he realizes that he is far too large to fit in his humble home.

The many corporations that drive our modern society are very much like monstrous Amik. They cannot ever be filled or satisfied. They are driven instead by an endless command to "grow" at all costs. Even if this means that our home—the Earth—becomes uninhabitable. Unlike ordinary people, corporations are compelled to generate increasing profits year after year through their corporate bylaws and boards of directors. They have no incentive to operate for the welfare of the public or the Earth. This has allowed the worst aspects of human nature, traits like greed and domination, to swallow up the Earth by way of the corporate "machine."

Governments are also often controlled by corporate owners to support their unsustainable actions, products, and technologies. This is done directly and indirectly by way of tax write-offs, start-up subsidies, and land grants. In the United States, the government hands over billions of tax dollars to "prop up" the research, exploration, and extraction activities of fossil fuel industries like oil, coal, and gas.[4] Subsidies for oil extraction and coal mining are so egregious that former President Harry S. Truman declared that he knew of "no tax loophole in the tax laws so inequitable as the excessive depletion exceptions enjoyed by oil and mining interests."[5] If taxpayers refused to subsidize these industries, they would cease to make money for their owners, who kept everything for themselves like the monstrous Amik. Instead, we would be implementing the many alternatives and sustainable ways to power our lives.

Industrialization promised us a safer world and a life of greater ease and comfort, yet it took away the most important things that human beings could want or have: the freedom to spend their time as they wished. This is because many "industrial jobs" brought people in from the outdoors to work in factories. In addition, it created a "mono-culture" of work and vocation. Expecting human beings to work for long hours indoors instead of achieving their individual dreams, or access their creative skills and gifts leads to social and spiritual impoverishment.

This disconnect between work and human needs, leads to individual and societal distress. Working non-stop is a fairly recent societal invention. Archeological records indicate that individuals in ancient communities spent time working together for the good of the whole and had a great deal of "free" time.[6] Most of these ancient peoples lived close to their families their entire lives, and most "work" was localized. This was especially true for hunter-gatherer communities and later agrarian communities.

We are social beings. Our well-being is largely defined by our relationships more than by money, or possessions. While having money, fame, and/or non-essential possessions is encouraged by a consumer society; long-term studies prove that nothing guarantees health, happiness, and longevity more than close relationships.[7] Unfortunately, the "improvements" in material conditions that come with "modern" civilizations have often led to the demise of intimate relationships; relationships with family, friends, community and other life forms. It's hard to understand others, if you do not spend the time.

According to Bloomberg Business Week, sales of outside storage units have doubled in just the past 15 years because people cannot fit everything in their homes.[8] In part, our ever-increasing desire for "more stuff" is fueled by corporate owners themselves. Having a populace that buys leads to increasing profits. Thus, the average "consumer" in industrial societies is exposed to 5,000 corporate ads a day.[9] It has gotten so bad that preschoolers recognize the brands and logos of companies like McDonald's, Coke, Pepsi, Disney, and Dannon. Yet they cannot recognize common animals, plants, and insects in their own backyard.

This literal marketing barrage convinces people to buy things that they do not want or need.

Jeanne Arnold, a member of a UCLA team studying modern American households, found that "*contemporary U.S. households have more possessions per household than any society in global history.*"[10] In the smallest home of 980 square feet, 2,260 individual items were recorded! It is literally impossible to keep track of all this stuff. Yet, this is not a problem that our ancestors had. Accumulation of thousands of "items" is a modern person's dilemma.

~

During the International Bering Sea Conference's planning stages, I was invited to stay with a local Tlingit family in Alaska. Tlingit means "*people of the tides.*" The family lived in a simple two-bedroom home in southeastern Alaska and came from a long lineage of fishing families. While their home looked in many ways very similar to a modern home, there were distinctive differences. For one, they had a beautiful blanket made from bark and goat hair draped over a large chest in their living room. I was told that the wife's great grandmother had made the blanket, and although it was slightly faded, rich colors of orange, ochres, and yellows still shone in the design. Besides this and some very humble basic furnishings, the house was largely unimpeded by other possessions or clutter.

The next day, the father, the wife's brother, their son, and two neighboring families built a traditional canoe, and I was invited to watch them as they worked. As I stepped outside into the cold and drew the warm coat around my neck, hard snow crackled under my feet. I walked along to the house's side, where a long trunk of a western red cedar tree was lying on its side. They had taken great care in selecting the tree to use for the canoe and had made some good progress in hollowing it out. After a few brief introductions, I stood on the packed icy snow to watch them while they worked. The four men spent time quietly talking, walking along the trunk's length, and pointing to different areas. While many of their words were in English, they interspersed their sentences with native words.

About twenty minutes later, the neighbors' 22-year-old son arrived. He approached with his head hung low, issuing soft apologies for being late. Joining the men, he stood with them in a small circle near the unfinished boat's bow. The youth was handsome, with a chiseled face that could be seen better from between his shoulder-length hair when he looked up, which he only did intermittently. Everyone stood quietly for a few minutes. Then the boy's father said in a low but calm voice, "*You have been drinking*." His son's face flushed. He looked up at his father briefly and nodded. "*You will come back tomorrow then, eh*." his father said, firmly but softly. The youth lifted his face, and his eyes opened wider. With a slight smile of understanding, he nodded and walked away.

After he left, the men went to work. They picked up their tools, chisels, and even three old-fashioned adzes made of stone and wood. One of the adzes had a whale carving on top, which I learned belonged to the father whose home I was visiting. He was of the Whale Clan, while his wife was of the Raven Clan. The men continued to advance the hollow's size in the log by chipping away methodically with their chisels and adzes. The adzes' synchronistic movements as they hit the wood, followed by their muffled sounds, had a calming effect. Yet, I continued to wonder why they sent the young man away; he may have been hung-over, but he seemed in total control of his body and appeared able to work.

At dinner that night, I asked the couple why the young man could not work on the canoe. My host explained that it was understood by all that he would not work on the canoe with a hangover from the night before. This was because the intentions and energy he could muster in that condition were not considered suitable for a seaworthy boat. It was understood that he could come back tomorrow and join them if he did not again drink. This was a powerful lesson for me about the importance of intention when producing objects by hand.

They believed that the intentions and energy of the maker become intermingled with the created object. Therefore, thoughtful handling and focused intention imbue objects with the best beneficial energy. The care brought to bear in making the object is a large part of what makes objects valuable and, in some cases, even sacred. In the case of the making of a canoe, first, the tree is honored when it is harvested, and this respect is maintained throughout the process of creating the canoe. Shamans of old understood this well and thus could imbue their sacred objects with supernatural energy.

Sacred objects are found in most belief systems. The "Shroud of Turin," considered the burial cloth of Jesus, the relics of the Passion in Notre Dame in Paris, and even the Bible used to swear an oath in court, are all examples of objects deemed to contain tremendous power. For Muslims, the Quran is considered the literal word of God and so sacred that it is not to be touched by anyone who is not "purified."[11] In Muslim homes, the book often sits on a pedestal stand or is left in a clean, uncluttered area. It is never allowed to be underneath anything else or to touch the floor.

I own a few objects that have been "made sacred" and contain powerful beneficial energies. One of my esteemed teachers, Sensei Hiraoka, gifted me with several objects that contain these. Sometimes they can be used to heal, other times simply to increase one's energy force. Other objects that I cherish were gifted by one of my most esteemed mentor and friend, Denise Linn. Her powerful soul-coaching program is based on Air, Water, Fire, and Earth and bears many similarities to the ancient Celts' ways. Denise gave me a beautiful singing bowl.

I use this bowl, which is named "*Singing the Animals Back to Life,*" to honor Nature. I take it with me as I lead quests and ceremonies worldwide. It has traveled with me to Hawaii and all the way to the Orkney Archipelago. During group ceremonies, participants "place" their energy and loving intentions for trees, the animals, and the Earth into the bowl. I can observe the healing energy filling the bowl. When the bowl is full, it is then "rung" with the group visualizing our thoughts, prayers, and intentions spreading throughout the world, carried by the vibrations of the bowl's song around the planet.

On the opposite end, the trillions of objects made carelessly and cheaply in mass production do not contain energetic power for their owners. In fact, many of these objects may actually pull down their owner's consciousness and energy. Our modern machine made "stuff" not only can "clog" up our homes and places of work; it can even "clog" our minds and hearts. Purchasing many items may keep the wheels of commerce turning, but it can engender a mediocre and unfulfilling existence by surrounding people with harmful energies.

For example, while a beautiful handmade quilt that was lovingly created by a grandmother or relative may bring joy, happiness, or comfort whenever it is seen or used, a machine-made quilt may provide neutral or even non-beneficial energy to its owner. The owner of such a quilt may have their energy diminished or their thoughts clouded when they see it or come into its presence.

I can transmute the energy of specific objects to make their energy beneficial to their owners, and or finely tune them to pick up desired subtle energies and vibrations. Material objects, while not "alive" can be influenced with one's intentions, prayers and sound. I have also helped people to transform energy of objects from being non-beneficial to shift to positive frequencies. This is a skill that can be easily learned.

It is possible to tune into an object's energy and understand if it brings your energy up or down. Sometimes though, it can be difficult to ascertain an object's influence, especially if one's thoughts about the object strongly interfere. For example, a friend may have gifted an object, or the object may be expensive and appear superficially grand or beautiful. In that case, it can be a challenge to discern whether the object is helping your energy or not. This is why it can be helpful to have another person trained in energy work to facilitate one's understanding about a personal object.

This is why I developed a remote dowsing technique to help people accurately identify an object's effect on their energy system or aura. The remote dowsing technique can bypass the conscious mind's limits to reach the subconscious, intuitive realms. Many people have benefited from this technique and have been able to pass along objects that bring negative vibes into their lives while accentuating the influence of objects that bring them positive vibes, thoughts, and emotions. This "dowsing method" can help people via phone or Skype anywhere in the world. The technique relies upon traditional dowsing materials like dowsing rods and pendulums, along with my body's own ability to sense subtle energies.

One of the most astute dowsers I have worked with is Joey Korn. Joey has been dowsing for almost 50 years, and has advanced the field of energetic dowsing considerably. According to Joey, "*For me, dowsing, or "divining," is not just about finding a place to drill or dig a well or about finding anything in the physical world. It's about detecting energy. It's about tapping into our own God-given ability to explore the world of subtle energies that are everywhere within and around us. It's about drawing ever closer to the Divine in our lives. That is why I like the term "divining."*"[12]

Societies that reflect cultural and economic paradigms to "get" rather than to "give" are using an artificial construct, not a natural one. Knowing that physical objects can affect their owners, energetically is very important. It is also important to understand that while some people may have a natural proclivity to accumulate things, it is not a natural human condition. Mass buying and consumerism is a societal, corporate mandate that has influenced people's psyches for generations. We can learn from indigenous societies that "objects" hold energy, which is why we need to take care of every object that we make, use, or own. Today, as people free themselves from the "priority" of buying and consuming, they will not only escape the negative influence of too much non-beneficial stuff but also gain the clarity, and focus to lead more meaningful lives.

11. Goodbye to the Trees

Cha bhi fios aire math an tobair gus an tràigh e.

Common Celtic Saying

In the fall of 2011, I flew to Austin, Texas, on a commercial jet to meet with two former board members of the Texas Audubon Society. I rarely flew myself now, and I was just beginning to feel comfortable flying as a passenger in large jets, versus piloting the more agile single-engine planes. Uncomfortable in the small cramped seat, I craned my neck to look out the tiny window. Frustration got the better of me, as I whispered to myself, "*The bigger they are, the harder they fall,*" referring to the great size of the jet at 700,000 pounds. As the word "fall" left my lips, it triggered a memory of one of my very first flying experiences.

One beautiful sunny morning, when I was still a student pilot, I had taken the plane up for a third time with my flight instructor, Rob. Rob was tall with pale-blue eyes, freckles, and dark, reddish-brown hair. He always acted older than his 28 years and was always well-dressed in clothes that gave a neat "uniform-like" appearance. He was an excellent instructor and a pleasant flying companion, with a relaxed but disciplined teaching method: no easy feat when working with people learning to fly thousands of feet in the air. After developing the flight plan, I boarded the plane and sat in the pilot seat while Rob settled into the co-pilot seat.

There was no crosswind at takeoff, so we lifted gently off the tarmac into a clear blue sky. For thirty minutes or more during our flight, Rob and I chatted about the weather between identifying sites below for navigation. It seemed to be a routine flight. I was relaxed and alert as I kept tabs on the dials on the control panel as well as our altitude and speed. Once we reached our optimum altitude of 14,000 feet, Rob fell silent as if deep in thought. I continued to fly and planned to stay at this altitude for at least another 45 minutes before descending for landing. I looked out the windshield to check our location against land features to make sure we were on track in our flight plan, when all of a sudden, Rob leaned forward and grabbed the keys out of the ignition.

In an instant, the 4,000-pound plane turned from being a graceful "steed" of the air to a hunk of weighty metal. A thousand thoughts began to fill my mind, including whether Rob had lost his mind and had a death wish. Yet, I was so stunned I was unable to speak. The plane shuddered and began to drop rapidly from the sky. As it made its terrifying descent downwards, the nose of the plane began to spiral. The sudden drop in altitude made it feel like my stomach had relocated to my throat. Around and around we went, as the plane dropped in tight dizzying spirals towards the Earth below.

As we fell hundreds of feet in seconds, the blood rushed from my head. I was close to fainting. Believing that our very survival was at stake, I somehow pulled myself together and forced myself to stay conscious and alert. As the seconds ticked away and the ground came closer and closer, I leaned forward and grabbed the yoke.

I pulled back on the yoke with all my strength until the nose of the plane was lifted. Once the nose was parallel to the ground, the dynamics of "lift" began to take hold. The disorienting nose-down circling stopped. The plane set into a gliding perspective. We continued to lose altitude, but at a much slower rate, and now we were also moving forward instead of straight down. As the plane glided, I started to do calculations in my head about how many minutes we might be able to maintain our glide before hitting the ground. I looked around for a safe, somewhat flat, and open place to land. I was ready to use a field or a quiet road if needed, when Rob, who had not said a word or even moved during the ordeal, leaned forward and put the key back in the ignition. The plane's engine started. I was never so happy to hear that loud mechanical roar.

As soon as I recovered from the joy and relief of having the engine back on, I glared questioningly at Rob. Without looking at me, avoiding my angry eyes, he abruptly blurted out, "*Excellent job.*" I looked at him as if he were out of his mind. While still looking ahead out of the cockpit window, he explained that he did this "maneuver" to all new pilots in training. "*Essentially,*" he said sheepishly, "*if you passed out, panicked, or could not handle losing your engine power, your flying days would be over.*" As this explanation began to settle on my agitated mind, he added, "*I know it can be a bit scary, but there is no point in me telling anyone of this test beforehand because it's supposed to be a surprise, and, after all, you did very well, so you should be proud.*"

It took a few days, but eventually, I did feel proud. I had been presented with a life-threatening emergency and took actions that could have saved us had the emergency been real. At the same time, I was also aware of the initial and debilitating effects of shock and disbelief on my ability to do the right thing in the short amount of time we had. That lost us important time when we needed every moment to recover. When one is faced with an unforeseen and unexpected danger, inexperience or delay can impede survival.

Just as this thought came to mind, as I peered out of the tiny window of the commercial jet now coming in low over the Texas Hill Country towards the airport, my heart skipped a beat. I saw, to my amazement and dismay, not the woodlands and trees that usually dotted this landscape, but instead thousands of tree skeletons with leafless, deadened branches stretching forlornly towards the sky as far as the eye could see. The sinking feeling caused by this sight stayed with me after the plane landed.

When my colleague from the local environmental non-profit picked me up at the airport, he told me that millions of native Texas trees had died that very summer. These trees, such as southern wild oak and bur oak, are perfectly suited to this hot, arid part of the world, and yet could not handle the excessive heat and drought presented by global warming. Later, during meetings downtown, local ranchers and scientists told me that over half a billion trees had died that year alone. No one could recall a time when this had ever happened before, nor could their parents nor their grandparents. There was simply no known case of such a massive tree die-off in recorded history. When I returned home a week later, I knew that the California drought was not an isolated incident. In fact, it is not just Texas and California experiencing massive tree die-offs; tree die offs have been documented on all continents, except Antarctica.

For example, the sacred Hawaiian Ohi'a tree, a flowering evergreen that covers 80% of the state's canopy, that sustains birds and insects found nowhere else in the world, are being wiped out by a dangerous fungus. In Canada, mountain pine beetles are literally eating through entire forests. In 2015 alone, the beetles felled over 730 million cubic meters of pine trees in British Columbia. B.C. is a major exporter of wood to the U.S. and its expected that the beetles alone will be responsible for destroying 55% of marketable pine trees by 2020. Eucalyptus trees in Australia are also in severe decline from cutting for agricultural purposes, pathogens, fire, and pollution. It is estimated that globally these trees have declined by no less than 30%.[1]

Trees today face a veritable soup of hazards. Entire forests in several states and across the globe are being killed off from drought, heat, and insect infestations. Bark beetles, which used to be controlled in the past by freezing weather and healthy trees able to flush these bugs out of their system with sap flows, are now on a rampage. With the warming climate, winters are warmer and milder, with less snow. In the spring there is now less rain. Trees too weak and dry to flush the bugs out with rising sap in the spring cannot repel the beetles as they once did. Bark beetle infestations are now the largest and most severe in recorded history.[2]

In extreme heat, even rain does little good to moisturize the soil. Just as damp soil in a hot oven dries out quickly, rainfall simply evaporates away in exceedingly hot weather leaving the soil dry. Then there is simply not enough soil moisture left to support the roots of plants. Trees cannot handle the increased heat and unusual precipitation patterns global warming brings. In contrast, pests that harm trees, like the pine beetle, seem to thrive in the Earth's warmer, unnatural condition.

An imaging spectrometer able to use sunlight to measure a tree's water content proves that even "green appearing" trees may be drought-stressed beyond their limits and dead within the next 6-12 months. According to the spectrometer, millions of trees show low amounts of water and are at risk of dying unless weather patterns improve. This is because, during a drought, the root systems of trees are damaged. They are then less able to access the water they need by way of their roots and are also subject to bending and falling over. The spectrometer is stationed onboard the Carnegie Airborne Observatory flying over forest lands in California to help predict where the next major die-offs will occur.

Even the long-lived Ponderosa Pine, Pinus Ponderosa that is thousands of years old and has survived all kinds of environmental changes, is in trouble. Forest Service professionals believe that these iconic trees will all be gone in just 10-20 years.

Dry trees become like standing wood logs that are highly susceptible to fire. Forests filled with dead trees and dry trees are literally tinderboxes for forest fires. Unlike fires of the past that allowed burned habitat to recover in the following years, the roasting fires change the region's biodiversity and keep native plant species from propagating by turning their seeds and acorns into nothing more than ash and smoke. The fires heat the local air system, and the flames reach so high that they create their own weather system. These fires are nothing short of "apocalyptic." This is why the fires burn so hot, with flames reaching as high as 45,000 feet![3] Firefighters are finding it impossible to keep up. It is estimated that in California alone, over 150 million trees have perished from climate changes, directly or indirectly.[4]

Studies conducted by forest management agencies estimate that more than 147 million trees have died over 9.7 million acres since the latest extreme drought period started in 2010.[5] That is nearly a third of the state's total forested terrain (33 million acres)![6] Called the "*worst epidemic of tree mortality in modern history*" by former Governor Jerry Brown, this statewide tree die-off is already causing a dramatic loss in biodiversity.[7] Plants and trees are the foundation for other life forms, not to mention the providers of half of the oxygen on the planet.[8]

For example, the Amazon rainforest contains 20% of the worlds biodiversity and its forest provides rivers of atmospheric water that are critical to life on all parts of the globe. From 1970 to 2012, 20% of the forest was cut or burned primarily for cattle ranching. Laws were passed and fines applied to slow the demise, however under the new President and former military leader Jair Bolsonaro, those laws have been slashed.

As Bolsonaro opens the rain forest to unmitigated destruction, he not only harms regional biodiversity, he is endangering moisture and weather systems that affect the entire world.

In the United States where fires have overtaken the west with a fierceness almost incomprehensible, old school and ineffective forest management practices continue to be applied. For example, forest thinning, slashing, burning and application of pesticides continue to be used by traditional foresters yet these are not an answer to stopping forest fires or dying trees. In fact, forest cutting and thinning has proven to make forests susceptible to hotter and faster burning fires. This is due in part to the increased air and wind circulation that occurs in managed forests versus those that have denser plant biomass. In addition, thinning forests also reduces transpiration of the canopy and causes drier conditions.[9] Trees are masters at moving water from the ground up into the atmosphere. Thus, trees and plants create a moisture rich atmospheric conditions in the region in which they are found. Massive forest systems, like those found in the Amazon, put water into the atmosphere forming global rivers in the sky.

On a single day, the rain forest produces 20 billion metric tons of water that is released into the air, and travels on currents of air flow in the atmosphere. To put things in perspective, this "invisible river" "manages" as much water as almost 50,000 Itaipu Dams. The Itaipu is the second largest hydroelectric plant in the world. This is because each tree that brings water into its body, only uses about 5% for its life systems, the vast majority is released into the air via a process of transpiration.

Trees equal water. They keep moisture and water in the near environment, and in the case of great forests like the Amazon, they drive a far reaching global water and moisture system. The atmospheric river created by the forest waters vast areas in south America keeping the land lush and green. The atmospheric river also creates worldwide weather patterns that all life depends upon and brings nutrients to oceans to feed microscopic life that the marine biodiversity depends upon for survival.

Thus, deforestation is not just a "backyard" problem, it is a global one. In addition, as trees die off, they are no longer able to take carbon dioxide from the atmosphere, which leads to further global warming.[10] The smoke and ash caused by wildfires can also cover glaciers and snow, causing them to darken and then melt more quickly in the sun.[11] Tree-loss in one region, like the Western United States, can affect weather patterns on the other side of the country. According to a National Science Foundation study, forest loss is disrupting the flow patterns in the atmosphere over parts of America and beyond.[12]

Daniel Griffin, a dendrochronologist at the University of Minnesota, has further evidence that California's drought was highly unusual. Griffin studied Blue Oak tree rings to determine how exceptional the California drought was in relation to other droughts in the past. Blue Oak trees store data in their rings that indicate soil moisture and rain patterns. Each ring is an accurate historical record of a tree's growth. The trees' data is far more extensive and accurate than rain gauges and manmade devices, which have only been in use for 75 years. Not only do tree rings provide much older data (even previously cut or preserved trees still have rings that can be read), but their data reveals what matters most: the impact of drought on plants and living things. Griffin was shocked to discover that the trees tell us that the recent drought was the worst in 1,200 years!

The Blue Oak Study, published in a partnership with Woods Hole Oceanographic Institution, came to the conclusion that:

"We're moving into levels of carbon dioxide concentration in the atmosphere that have not been experienced any time during human history, and probably for the last 3.5 million years or more. Also, this type of hot drought, where you've got low precipitation magnified by record high temperatures, is a pretty good prototype for the type of drought that we expect to see in the twenty-first century as temperatures continue to rise in direct response to human emissions of greenhouse gases into the atmosphere."[13]

One of the last civilizations to experience a drought similar to ones we experience now was the Mayan civilization in Central and South America. In 850 B.C.E., the Mayans' Yucatan territories were subjected to extreme drought conditions that lasted on and off for over 95 years. Archeologists now largely agree that the drought almost wiped out the entire civilization and brought a once vibrant civilization to its knees.[14] Natural global climatic changes influenced the reduction in rainfall, but it was the Mayans' extensive clear-cutting around their largest cities that exasperated unusual weather patterns, leading to severe long lasting drought. These droughts led to upheavals among indigenous civilizations in the Southwest as well.[15] A recent study by Columbus University published in Science explains that California has entered into a megadrought period, which will have catastrophic impacts on the region.[16]

~

Once sanctuaries, where I felt enlivened and at peace, the hills and valleys I frequent are now sad indicators of the devastating effects of global warming. Declines in animal, bird, fish, and reptile populations are noticeable. Bird species have suffered a precipitous decline all over the world. Like the northern pintail, horned lark, and loggerhead shrike, some species have declined by as much as 75 percent and are just clinging onto existence.[17] The evening grosbeak has declined by a staggering 96 percent in only the past 10 years. The California whitetail deer population has also declined, 60 percent, and likely more, in the last 20 years.[18] Small mammals, like voles and deer mice, and are also plummeting. Because these animals serve as the foundational food source for raptors like owls and hawks and larger mammals like coyotes, mountain lions, and bobcats, their decline puts the whole food chain at risk.

While some people are aware of the major declines in animal species and trees in their communities, most are not! This is because they spend so little time outdoors. Those who spend time outside, and pay attention to the wild things, or study Nature are seeing losses similar to mass extinction events like the Cretaceous-Paleocene extinction that occurred 66 million years ago and wiped out three-quarters of plant and animal species. Of course, there were no humans to experience that event, so the rapid disappearance of plants and animals today is not only extremely rare but unimaginable to most people. Global warming is the human equivalent of a deer in headlights. However, we must quickly get over the shock and take action to avoid catastrophe.

Just thirty years ago, during the 1990s, there was a more cohesive understanding within the majority of the human population about climate change and its impact on the planet. Unfortunately, as public awareness began to take hold, corporate owners in the oil and gas industries recognized that people were not only getting wise to this disaster but were beginning to pinpoint those largely responsible for it. Exxon Mobile was aware that its "business model" would lead to devastating climate change as early as 1977. According to investigations by "Inside Climate News" and Harvard University, this was 11 years before global warming became a public issue.[19] Yet, instead of revealing this information that affects all life on the planet, the company not only refused to go public, but it did a deep dive into obfuscation of the truth.

The company spent thousands of dollars promoting climate change denial and misinformation. This was the same unethical approach used by the tobacco industry to lie about smoking's health risks. Many more people developed cancer and died from smoking because tobacco companies obstructed the truth. Unfortunately, today, oil company led propaganda campaigns have blurred the truth about climate change and its causes. This has directly resulted in today in public confusion and inaction. Unlike smoking, where it was easier to directly link individuals smoking habits to their illnesses, understanding a global problem is a challenge for a lay person, especially one who is not aware of changes in weather, and wildlife.

In 1989 Exxon Mobile led the creation of the infamous "Global Climate Coalition", which starkly questioned the authentic science behind global warming. This "Coalition," a disguised industry-led lobbying group, prevented the US from signing the Kyoto Protocol in 1998, which sought to limit greenhouse gases. This diabolical tactic was responsible for stopping other countries such as China and India, as well, from signing the treaty.[20]

This all came to light during an eight-month-long, in-depth investigation of former Exxon employees, scientists, and federal officials. The investigation unearthed Exxon's senior scientist, James Black's sobering message to Exxon's management committee: "*In the first place, there is general scientific agreement that the most likely manner in which mankind is influencing the global climate is through carbon dioxide release from the burning of fossil fuels.*"[21] He warned them that doubling CO2 gases in the atmosphere would increase average global temperatures by two or three degrees—a number consistent with the scientific consensus! Black continued to warn them that: "*present thinking holds that man has a time window of five to 10 years before the need for hard decisions regarding changes in energy strategies might become critical."*

It's hard to believe that these corporate leaders chose to protect their profit margin over the welfare of life on Earth. People who make such decisions would ordinarily be deemed "insane" in a healthy, sustainable society. Today too many people, especially in America and in some other Western nations, actually believe either that planetary global warming does not exist or that it is actually good for the planet? These beliefs are[22] largely responsible for defeating national and international measures to limit carbon releases. It's devastating that a "win" for a corporation like an oil company is considered an acceptable "loss" for the world. These oil company executives are making the world unsafe for all life, including their own children, grandchildren and descendants.

Today, I am saddened and angry to hear people say confidently that "*global climate change is naturally caused.*" They argue that humans have little to do with the warming trend and that levels of CO2 in the atmosphere, a harbinger of climate and temperature changes, have always gone up and down. They may not be aware that they are, in fact, spouting one of the major lies developed and spread by Exxon. However, real science tells a different story; climate change and CO2 levels prove that it is human-caused and dangerous. For example, there are three distinct types of carbon isotopes: 14C, 13C, and 12C. The isotope associated with carbon from burning fossil fuels has risen dramatically since the 1900s; meanwhile, the naturally derived isotope associated with carbon created by trees has gone down.[23]

In fact, sequences of annual tree rings going back thousands of years indicate that at no time in the last 10,000 years have the naturally occurring carbon isotope ratios in the atmosphere been as low as they are today.[24] Not only are naturally occurring carbon isotopes falling, but so is Earth's atmosphere's oxygen level. Oxygen levels are falling 2,000 times faster over the past 150 years alone than in the last 800,000 years.[25]

Greenhouse gases absorb energy and will thus allow less infrared radiation to escape into space. This is exactly what satellites are finding. From 1970 to 1996, satellites discovered that less energy is escaping into space and proved via "*direct experimental evidence of a significant increase in the Earth's greenhouse effect.*"[26] Moreover, if global warming were predominantly caused by forces outside of the planet—for example, by a hotter sun, another oil industry-created myth—then the outer layers of Earth's atmosphere would be heating up as well. However, this is not the case:

"Climate models predict that more carbon dioxide should cause warming in the troposphere but cooling in the stratosphere. This is because the increased "blanketing" effect in the troposphere holds in more heat, allowing less to reach the stratosphere. This would contrast with the expected effect if global warming were caused by the sun, causing warming both in the troposphere and stratosphere. Instead, satellites and weather balloons confirm a cooling stratosphere and warming troposphere, consistent with increased carbon dioxide."[27]

Some "climate change deniers" will rely on ice samples taken from the Vostok core at the Earth's polar region to support their argument that the Earth's present warming trend is a naturally occurring cycle and thus ok? They refer to an ice-drilling project between Russia, the United States, and France at the Russian Vostok station in East Antarctica. This yielded the deepest ice core ever recovered, with layers of ice going back 800,000 years.[28]

While the drilled ice samples show that the Earth has gone through cycles of warming and cooling trends, there have not been periods in Earth's history when temperatures were warmer than they are now. Also, the rate of climate change today is far more rapid than the climate shifts that occurred in the past. The Vostok ice cores show us that the rate of change in CO_2 over the past million years is tame compared to today's rate. Before the Industrial Revolution, CO2 changed less than 0.15 ppm per year.[29] Today's rate of change is twenty times faster!

Even if climate change were a "natural occurrence", it's still a forerunner to mass extinction and will undoubtedly impact the human race. For example, in the end-Triassic extinction 200 million years ago, CO2 doubled from 2,000 to 4,400 ppm, triggered by massive volcanic eruptions.[30] This happened over the relatively short time period of 1,000 to 20,000 years and wiped out 75 percent of all land species and 95 percent of all marine species. If humankind had been around at that time, we would also have become extinct! If we compare this event to our present global warming, the rate of change that occurred then is literally just a fraction of the rate of change occurring now. In April 2020, our planet had 419 ppm of CO2 in the atmosphere, and this is rising rapidly! (For updates, go to https://climate.nasa.gov/vital-signs/carbon-dioxide/.)

Already we are observing the beginning of end-Triassic conditions. In that period, oceans became acidic soups devoid of oxygen, suffocating almost all marine life. Unfortunately, dead zones are now appearing in many different locations in our oceans.[31] Extinction events are not something that changes back to safe levels quickly. The end-Triassic extinction showed that it could take millions of years for life to recover and for new species to evolve. It is estimated that it took between five to 30 million years for life to begin to reappear after that mass extinction.

For those that continue to ignore the realities of global warming and its causes and damage, a final but important point is that the fossil fuel industry is outdated. It lags far behind new, more sustainable, and less expensive technology. In fact, the design for the combustion engine used in most vehicles is over 100 years old! Moreover, the oil and gas industries pollute the land and water through their pipeline and tanker oil spills and fracking. By inventing newer and better alternatives to generate energy and investing in them with our tax dollars, we can do better. It's time to let go of these "dinosaur" technologies and apply the new methods that run far more efficiently and safely.

As if historical records were not dire warnings enough about the impact of increased CO2, we have also to add two other very important trends that are affecting Earth today. These trends were not taking place during the period of the Triassic. One, we have allowed massive deforestation from clear-cutting, forest fires, and soil and water contamination. Over the past 200 years, humanity has wiped out approximately 50 percent of the Earth's forest cover.[32] Thus, not only are we pumping carbon dioxide into the atmosphere at an alarming rate, but we are also wiping out the very species we need to help regulate our climate. The reality is that not only has humanity never faced this type of rapid and devastating change under these dire circumstances, neither has the Earth.

As I look at charts showing the rapid rise of CO2 and corresponding temperature increases worldwide, it reminds me of a falling plane without an engine. We must take action or leave the world spinning into an uncontrollable collapse. In 2002 world leaders set targets to conserve key habitats, stem pollution, and conduct important ecological research. Yet, just eight years later, in 2010, the Convention on Biological Diversity published a Global Biodiversity Outlook that contained a grim update. The convention's report concluded that:

"The loss of biodiversity is an issue of profound concern for its own sake. Biodiversity also underpins the functioning of ecosystems, which provide a wide range of services to human societies. Its continued loss, therefore, has major implications for current and future human well-being."[33]

This report is a wake-up call and a reason to take action. This is not the time to sit back and think that someone else will save the world. Whether this new era ends well or not depends entirely on us. It is simply not possible to run away from climate change by ignoring it. Nor is it possible to "leave the scene." There is only one Earth.

~

An article titled "Eco-Anxiety Takes a Toll on Global Warming Alarmists" claimed that "Eco-anxious people" *have unrealistic, neurotic fears that can and should be "cured,"*[34] Another article in San Francisco Magazine, "Green with Worry," proposes that some people are what the author termed *"Eco-neurotic."*[35] To prove this point, the author cites the example of a forty-year-old professional woman, Audrey. Ever since Audrey heard a BBC report a few days earlier, she has not been able to forget the image of dozens of bears floating in the Arctic waters, dead from exhaustion after trying to swim to solid ice that no longer exists. Audrey then breaks down in tears in her therapist's office. *"Polar bears aren't supposed to drown,"* she tells the therapist, crying.

Audrey's dilemma reflects a growing trend. The term "Eco-anxiety" is used to label anyone who experiences feelings of helplessness, fear, anger, or despair over the state of the planet. However, grouping all people concerned about the environment as suffering from a "disorder" makes it seem like they need to be "set right."

Some psychologists and therapists claiming to be "experts" in the field of "Eco-anxiety" offer similar solutions: "patients" are to avoid or ignore reports and facts related to species loss, global warming, and pollution. Instead, they are instructed to partake in "feel-good" activities like listening to peaceful music or taking a warm bath. Yet these "cures" are nothing more than band-aid solutions to nullify or hide a real problem that needs attending to. The immense sadness and pain of coming to terms with the destruction of the world and the extinction of species is an Earth-sized sickness, a planet-sized mourning. Each one of us is like that cursed ancient mariner recoiling in abject anguish because he shot the albatross:

God save thee, ancient Mariner!
From the fiends, that plagues thee thus!
Why look 'st thou so?
With my crossbow
I shot the Albatross.
Since then, at an uncertain hour,
That agony returns;
And till my ghastly tale is told,
This heart within me burns.[36]

So what can someone do who is struggling with "Eco-anxiety"? We must do more than take a warm bath and listen to soothing music (although these are good temporary treatments to gain an emotional equilibrium and getting professional support may also be needed if one is chronically depressed or worse). Coming together as a people to share our sense of loss and hurt is a start. Informing and educating others about the truth that global warming exists as well as sharing personal stories of loss caused by it.

These stories can help us unite into a shared understanding of what is really happening in our world. We must also take action. Supporting and electing political candidates that platform a green agenda, reign in fossil fuels, and supporting alternative energy programs, is a requisite for our survival. Putting unethical and harmful corporations out of business is also a must. We already have many alternative solutions to power our world. Many of these are crushed by corporations to maintain monopolies. We must unleash these new technologies and start heavily subsidizing them to get them out in the market-place.

At an individual and personal level, we can also take actions in our own lives such as not investing in oil and other extractive industries. We can also help by caring for the native trees and wildlife in our own backyards and neighborhoods. We can leave out water for animals and birds, and provide native plants that provide food, for example. There are also many wonderful non-profit organizations to support either financially or as a volunteer lending a helping hand. One of my all-time favorite environmental programs is STRAW, short for "Students and Teachers Restoring a Watershed."[37] I became familiar with STRAW over a decade ago and attended one of their restoration days. I was asked to lead a peer-to-peer evaluation of their work.

STRAW started small but has grown impressively. It is an education and restoration program that has engaged over 46,000 students in over 550 restorations on creeks and wetlands. To date, young people eight years old and up have planted over 50,000 native plants, restoring over 33 miles of habitat. Each year, STRAW conducts 45 to 50 restoration projects that include planting native plant species and restoring stream beds. One of STRAW's most important achievements is the quality of their work. Children who may have never before held a worm in their hands or spent much time outside are completing restoration projects that rival professional companies' work. These projects improve water quality and create healthy new habitats for wildlife, making way for the return of songbirds, fish, and other native species.

The project took form after a fourth-grader asked a simple question. Students were understandably crestfallen after their teacher Laurette Rogers showed a National Geographic film on rainforest destruction. The students' concern about the endangered species that depended upon the now-destroyed forests turned into energized action when in one pivotal moment, John Elliot, a ten-year-old, raised his hand and asked, "But what can we do?"

That simple question, innocent yet profound, ignited something in Laurette that launched her and her class to develop a program that led to the local environment's eventual transformation. *"I looked into his eyes," said Rogers, "and somehow, I just couldn't give him a pat answer about letter-writing and making donations.*" Rogers supported her students in creating STRAW, which allowed them to be positive agents of change. They restored damaged and neglected stream-side habitats and provided a safe home for an aquatic species on the verge of disappearing. STRAW continues to offer productive activities for young people to restore the planet. In return, it offers them a deeper connection to the natural world. STRAW is about habitat restoration, but it is also about humans healing their relationship with the Earth.

After a day of working on the land in STRAW's closing circle, the children close their eyes and imagine what the world will be like in 15 to 20 years. The STRAW project has been going strong now for over two decades. Those first students in Rogers' class are now in their thirties. When they return to the places where they first carefully planted native trees and shrubs, they now can see a nest of baby birds in one of the trees that sprung from a tiny sapling they once held in their own hands.

STRAW is a real "cure" for those suffering from "Eco-anxiety." It offers a healing antidote by involving people, young and old, in creating a better world. It is hopeful activism and one example of how human beings can partner with Nature to heal the planet.

12. A Calling

Dream lofty dreams, and as you dream so, you shall become. Vision is the promise of what you shall one day be; your ideal is the prophecy of what you shall at last unveil.[1]

James Allen

There were many twists and turns in my life before I chose to follow a path with a heart. There were signs along the way to return to the natural world and to live more deeply in the sacred. But without having grown up in a culture that supported this way of life, it was hard to hold fast to childhood beliefs and experiences. There were times chasing a "success" defined by society, I pushed myself into conforming to other people's notions. Yet, even then, I would get signs from the Universe coaxing me to take the road less traveled. A road that my heart longed for, even if my mind was not yet ready for the journey.

One of those signs occurred when I was only 28 years old. I had purchased a Euro-Pass and flown to Europe to see the "world." After spending time with friends and family in Scotland, I took a train to visit one of my mother's oldest friends in France. Jill invited me to visit her at her condo in Agde, one of France's oldest towns (founded in 525 B.C.E.). Jill was sixty years "young", and had started scuba diving in her fifties. She had gone on to become one of the best female divers in the world. When she picked me up at the train station, I was struck at how fit she looked. She had kept her beautiful figure and was wearing her reddish-brown hair to her shoulder, reminiscent of her appearance in my mother's photos of her from 30 years earlier. I admired how Jill had committed to following her passion after raising her family.

Jill enjoyed touring and hiking as well, so she showed me some beautiful places not far from where she lived. One day we visited Cap d'Agde, where Van Gogh stayed and painted his stunning sunflowers. Sunny fields filled with these bold golden beauties still exist. After staying with Jill for the better part of a week, I rented a car and took a road trip, an hour and a half north, to see the largest river delta in Europe, known as the Camargue. The Camargue Delta covers 360 square miles filled with marshland, wetlands, lagoons, and several lakes. It is home to a tremendous diversity of species, including over 400 species of birds.

I parked at the entrance and walked through a small woodland of ash, willow, and white poplar trees to get to the marshlands. Beyond the trees was a clearing of flowering yellow shrubs that thinned out to reveal a marshy plain filled with sea lavender. The sea lavender was in full bloom, and plum-colored flowers carpeted the ground with their petals. I ambled along, stopping now and then to enjoy the delicate scents of the flowers. At one point, I kneeled to get a closer look at the delicate flowers of the sea lavender when a large, bright-red insect flew past into the reeds. The creature's size and the dazzling color was stunning. I ran to get a better look beyond the reeds. There, perched on a mudflat, was a large red dragonfly sparkling like a living ruby.

As I admired its dazzling body, a loud, splashing noise caught me by surprise. Startled, I stood up to see six white horses running at full speed, knee-deep in the water not more than 200 feet from where I stood. The majestic horses moved in unison as gem-like droplets of water splashed around their legs. Their flowing manes and tails were illuminated by the sun. Their gleaming manes cast the light forward in rainbow rays. I felt as if I had been transported to another realm. The appearance of these majestic beings opened the door to a higher level of consciousness with their light.

After a few minutes, the horses passed on their way to a sandy shoal further down the coast, leaving me transfixed and staring longingly after them. When I drove back to Agde, I learned from Jill (this was way before Google and cell phones) that the horses are known as "the horses of the Camargue." They are descendants of the now-extinct Solutré horse of the Paleolithic era. The Camargue horses still have the distinctive large square heads and strong, stocky bodies of the Solutré horse. Some Camargue horses are captured and domesticated for use by the local cowboys to round up cattle. The horses are appreciated for their agility, love of water, calm temperament, and ruggedness. They can be ridden for long distances without tiring or injury.

Later that evening, I shared with Jill the uplifting vision of the horses. Jill nodded and shared some of the times that she, too, had had glimpses of another world elevated in a sacred dimension. One of those times she shared was of a spiritual experience that she had at Chartres Cathedral. Her story and the white horses inspired me to change my travel plans and go the very next day to Chartres to see the cathedral. The cathedral, also known as the Cathedral of Our Lady of Chartres, is considered a masterpiece of French Gothic Art. It was built between 1194 and 1220 C.E. Yet, its location, sited above a sacred spring, had been visited by spiritual seekers for many thousands of years prior.[2] The cathedral is said to reflect the three faces of the divine feminine. It houses an important sacred relic: Mary's cloth when she gave birth to the infant Jesus.

Jill dropped me off at the train station in Agde in the morning, and I used my Euro-Pass to get to the town of Chartres. I arrived in the late afternoon and walked from the train station into the old quarter of the city where the cathedral is located. As I walked uphill towards the cathedral, the sight of its spire caused shivers to run up my spine. I arrived around closing time, so I walked past to go to my bed-and-breakfast six blocks away. As I went to sleep that night, images of the white horses and the cathedral intermingled in my mind. In the morning, I got up early and then wandered around the charming hillside town before arriving at the cathedral doors.

While standing quietly, waiting for the doors to open, I observed the exquisite statuary above them. On the far left of the doors, there was what appeared to be an "unfinished" Christ; where he dwelled in a place before time, in the land of angels, before his incarnation on Earth. The statue was purposely left unfinished, as the artist sought to convey a spiritual being before coming into a manifest body. On the right portal was a statue of the Virgin Mary seated with the Christ Child on her lap. The ancient statuary presented a very empowering story of Mary.

The 12th-century sculptures represented Mary as a primal player in the divine story. Mary is not the "passive recipient," but the One through which a sacred being came into existence. Mary was thus the bearer of sacred potentiality. Just as matter is born from energy, and potentiality for all existence resides in the cosmic cauldron, Mary's womb was the divine cauldron from which a sacred being would arise. The statues made it clear that the sacred could not be born on Earth without the divine feminine.

Yet, in 1484, just 200 years after the completion of Chartres and statue, Mary's story as a powerful and divine being was to come to an abrupt change in many countries in the world. Pope Innocent VIII issued a papal bull, or document, that condemned "witchcraft."[3] The Roman Catholic Church hired a "police force" of inquisitors to seek out and bring "witches" to "justice." The Church also commissioned the Malleus Maleficarum, also known as the "*The Hammer of Witches*," one of the most vicious and evil books in human history.[4] The Hammer was used as the foundational text condoning the killing of tens of thousands of women, while millions of other women suffered torture, arrest, interrogation, hate, guilt, and fear.

These crimes against humanity were organized by the Roman Catholic Church and carried out throughout Europe and eventually the United States over a 500-year span.[5] The Inquisition inflicted physical harm and struck at the root of women's spiritual power and identity. By wrongly accusing women of being evil, many of whom were simply using their healing and prophetic gifts, women's societal, legal, and religious roles shifted dramatically.

Known as the "burning times" in feminist literature, the real motivation behind the Inquisition was to ensure that the Church's religious doctrine would dominate the beliefs of society.[6] Even if these beliefs and actions were in direct contradiction to the known teachings of Jesus. Church leaders were likely well-aware, even then, that witchcraft was not really an issue, but they sought dominion over people's beliefs.

One way to achieve their goals was to remove women from the sphere of truth-sayers, spiritual leaders and healers.

The witch hunts are closely linked in time and place to regulation of medicine and the healing arts as a "professional field." Before the 13th century, it was mainly women who were the pharmacists, herbalists, and healers.[7] They generally used natural remedies and gentle, non-invasive healing practices techniques. Women were also the midwives who encouraged natural childbirth. As male authorities began to "regulate" the field of medicine, only men were deemed allowed legally to practice. If a female healer or midwife should try to "practice" in the "old ways" without the "authority" of the new medical "industry," they could be charged with a serious crime. Many of the procedures that the new male "doctors" used tended to be more invasive, along with the growing use of chemically manufactured compounds. No doubt, these more invasive procedures, and lab-created concoctions gave doctors more authority and control over a patient's treatment and health. This also fueled medicine's rising costs as they turned away from herbs that grew naturally.

Many of the Church's victims, were herbalists or healers like Agnes Sampson, who was burned at stake.[8] Or they were women who had intuitive and psychic abilities and/or healing powers of touch.[9] In some cases, they were simply unmarried women ("spinsters" or divorced) who owned their own property or business. Their property and wealth could then be taken by the Church as these "witches" were legally required to pay for their imprisonment, torture, and death. (It would be interesting to track how much of the Church's enormous landholdings today were obtained during the Inquisition.)

The Church hired inquisitors and paid them "bounty style" for every "witch" brought to trial. Thus, no woman was safe from persecution. A woman might be considered a political enemy or economic competitor.[10] In one documented case, a scorned lover accused the object of his affections of being a witch. The poor woman was burned at the stake on his word alone. His reasoning appears to have been that if he could not have her, no one could.

Some believe that 400,000 women and men were killed during the Inquisition (85 percent of victims were women, and 15 percent were men).[11] Yet, this number does not account for the millions of people traumatized; who suffered the loss of loved ones or lived in dire fear for their lives. Most women killed were between 50 to 60 years old, followed by women between 40 to 50 years old.[12] Once a woman was accused, it was literally impossible to prove innocence. By today's standards, women and men accused of witchcraft had no rights. There was no presumption of innocence, no juries, and no right to an attorney. Those accused of witchcraft were simply assumed guilty.

They would be questioned, deprived of sleep, and sometimes tortured to obtain a confession. If none were forthcoming, then the accused would be "tested." These "tests" had nothing to do with reality or the "crime" in question. For example, one acceptable "test" was to tie the accused's hands and feet and throw them into water. If they managed to float, they were considered witches and executed. If they did not float and even possibly drowned, they were held to be innocent.[13]

The daughters and sons of the accused could also be tortured in front of them, as this was found to be a quick way to get a confession.[14] Inquisitors were always seeking names of other women that they could accuse to get paid a bounty. Getting those names from someone already accused of being a "witch" was the best place to start. Mothers, aunts, nieces, sisters, and daughters were tortured and killed and asked to turn against each other and other women to save themselves.[15]

Fortunately, today the institution of the Catholic Church is embracing a more compassionate dogma. Today, under Pope Francis, many new egalitarian ideas are being expressed and adopted, a welcoming new direction for the Church.[16] While the Inquisition may seem far in our past, witches were still killed as late as the18th century with the death of Anna Goldi in 1782.[17] Anna was a beautiful woman working as a maid for a wealthy employer. He claimed that she was a witch, and she confessed. Later, she denied the confession, stating that it was given under duress of torture, yet she was still put to death. In 2007, she was exonerated by the Swiss Parliament, which found that her execution was a miscarriage of justice. It was revealed that her employer had lied. He simply wanted to get rid of her to stop her from revealing their affair to his wife.

This terrible time in human history in which no woman was safe burned a deep psychic scar into the DNA of all relatives and descendants of women and men wrongly accused of witchcraft. The "burning times" left a profound and lasting fear in the hearts of those who partook in, witnessed, or heard about these events, forging a vast intergenerational thread of remembrance. Yet, there we have the power to overcome these past influences.[18]

~

Epigenetics proves that emotions, thoughts, and experiences can alter human gene activity and health. Genes within your DNA, give instructions to cells in your body to make molecules called proteins. When activated, for example, genes can keep you healthy by strengthening your immune system or if they are negatively affected by your emotions or experiences, they may fail to provide the protection needed. Genes also carry instructions that determine features such as hair and eye color and how effectively your body may deter specific illnesses like cancer and mental illness. Thus, one's state of mind and one's experiences can directly affect the activation of your genes—and this can affect your future emotional, physiological and physical state.

A truly remarkable study went further to show that epigenetic predilections can even be passed down to offspring and descendants. Physical, emotional, and behavioral traits can be passed from parent to offspring, based on one's parents' mental, emotional and physical experiences and states. Thus, ancestral trauma from a historical event like the Inquisition, or the holocaust could indeed have settled into the fabric of one's DNA and passed along to one's descendants.[19] While there are socio-economic and political forces at work that can hold people back from in their lives, perceptions and habits may be lodged as "genetic memories" from one's ancestors.

Yet, each generation and every individual has an opportunity to "erase" and heal not only their own past but the past of their ancestors. This truth has been known for centuries by those that continue to honor and work with ancestral energy. This reversal can take place with behavioral changes, environmental conditions, nutritional support, and substituting new beliefs for outdated ones.[20]

Each of us has the power to change our own "epigenetic inclination." According to Dawson Church, a researcher in the new science of epigenetics and the author of a "*Genie in Your Genes: Epigenetic Medicine and the New Biology of Intention*,"[21] "*This research reminds us that we hold many of the levers of healing in our own hands. It makes us aware that it is not doctors, hospitals, acupuncturists, homeopaths, chiropractors, energy workers, or other health professionals who determine our sickness and health. They can tilt the balance, but not nearly as much as we can. As individuals, each of us creates a big chunk of our emotional and mental environment, thereby turning genes on and off in our cells. This opens up vast and exciting potential.*"

Church not only shares a large number of fascinating scientific studies about epigenetics, but he also provides techniques and activities that one can do to change the proclivity of their own genes. We have the ability to stop passing non-beneficial influences to the next generation, forever freeing ourselves and our descendants from these negative patterns.

While each of us can take charge of healing our lives and our ancestors by removing past traumas, the Catholic Church could also redeem itself and admit that the Inquisition was a terrible mistake and a crime against humanity. It must further exonerate all of the women and men who were wrongly accused of witchcraft. By doing so, the Church enacts its own salvation and steps towards embracing a more compassionate, inclusive, and loving doctrine.

In addition, it is highly likely that "but for the Inquisition," the healing arts and medicine would have continued along a very different path. Today, 65 percent of doctors in the United States are men.[22] Similar gender inequalities in the top medical professions can be found in Europe and other countries worldwide.[23] Pharmaceutical companies, some of the largest and wealthiest corporations globally, are in the "business" of profit more than health. We need not throw out the "baby with the bathwater" and get rid of all "Westernized medicine," as many inroads have been obtained, but it is time to validate the more natural healing methods and the pioneers who bring them forth.

The healing arts are part of a woman's birthright. As women heal themselves from past trauma they step powerfully into the life-affirming roles that many are suited to fulfill. For example, the new "wellness movement" has roots going back to the 1970s and further was largely started by women. For example, therapeutic touch is a non-invasive healing modality that has proven benefits in inpatient care. Dora van Gelder Kunz and Dolores Krieger, RN, Ph.D., together developed this method, which has been used for physical, emotional, and spiritual healing. The underlying concept is that a "healer" can tap into universal beneficial energy and apply it to another person to bring them into balance, wholeness and health.[24] Not just women, but many men, are also now exploring and providing alternative, holistic and effective healing modalities.

For example, Dr. Tony Smith has developed a gentle natural method to cure one of the greatest threats to human health, Lyme Disease. Although not often in the news, Lyme disease has rapidly spread throughout many countries. It has been spurred on in part by global warming, making it easier for disease-laden ticks to survive in many locations that they normally could not. In fact, Lyme disease has many similarities to the recent Covid 19 pandemic. It is globally distributed and has led to over 500,000 of documented cases.[25]

Since Lyme disease is very misunderstood and mimics many other diseases, unlike Covid 19, it is often misdiagnosed. It is estimated that there are likely 2-3 times as many cases as those officially recorded. For example, Kris Kristofferson was originally diagnosed with Alzheimer's disease, but they discovered he actually had Lyme disease years later. In fact, studies have found the occurrence of Lyme disease bacteria in the brains of people who died of dementia.[26] Lyme disease is catastrophic to health and can cause life-long illness and even death. Unfortunately, Lyme dis-ease has been kept under the radar[27] and has received very little research, treatment, or attention and little funding to seek a cure. Yet, thousands of people continue to suffer from this debilitating disease.

After healing himself from Lyme's disease, Smith found that by helping one's own brain to identify, locate, and attack the disease, he could bring people back to full health and complete recovery. His technique is natural, highly innovative, and outside the traditional treatment protocols (which tend to be largely ineffective). If Smith had lived during the Inquisition, he would probably have been condemned as a witch. Instead, today he has healed and helped thousands of people to reclaim their lives. Smith is just one of an increasing number of medical professionals using alternative and natural methods.

Through the rising wellness movement, women and men have created an alternative health care system outside of, and despite, the mainstream medical establishment, which is just beginning to recognize the value of these traditional remedies. As they grow this new movement, not only are they creating a new way to health and wholeness for humanity, they are reawakening the ancient domain of women as healers.

~

As I stood on the street outside of Chartres Cathedral, waiting for the doors to open, a strange vision flickered before my eyes. The sounds of the bustling town vanished from my ears and were replaced by a loud rushing sound, like water in a powerful river. I found myself floating in the darkness of space with the planet Earth before me. The Earth appeared a dark, metallic grey, weighed down by some great heaviness. It was as if I saw an image of a tired, used-up, and forsaken planet. At the time, in the 1980s, I did not know anything about global warming or the planet-sized ills that were to become all too apparent in the next ten years.

Then the bright flash and a cosmic ribbon of white light came shooting out of the dark, primordial cosmos. The ribbon was several times wider than the Earth and so long that it was impossible to see where it began. The great band of cosmic light flashed brightly as it silently touched the planet. The affected part of the Earth immediately turned a milky white color. As I looked in astonishment and wonder, I heard a great and gentle voice say that this was a "softening." While this seemed a strange statement at the time, the voice sounded calm and wise. I took it to mean that some type of transmutation of the heavy metallic orb had taken place.

I then found myself no longer floating in space but flying above the ground of the planet at the distance of a single engine plane making a long distance flight. As I looked down, I eyed a village at the foot of a great mountain. I began to lower towards the village and was hovering at the same height of the mountain top above some very strange looking buildings when with a loud "clang" the large wooden cathedral doors opened. Startled, the "vision" slipped quickly away and I found myself standing at the entrance of the Cathedral.

I suddenly "came to" from the trancelike vision, and shuffled into the cathedral, along with a small group of tourists. As I entered the cavernous space, the slight tingling in my legs and hands that I had noticed earlier grew in strength. I walked to the middle of the cathedral and stood on a beautiful stone labyrinth. Energy radiated from the labyrinth and the ground beneath, making my body vibrate. The vision that I had received and the bodily sensations were linked somehow. I held my sides with my arms as if this would keep my body from vibrating. While I have experienced unusual sensations from being in powerful energetic places on the land, this was stronger than most.

~

Throughout the world, there are places on the planet that radiate strong, energetic forces. These special places in the land, where energies coalesce, were used as sacred sites by the ancients. The ancients sensed these earth energies and built their places of worship in these locations. Chartres Cathedral is a perfect example of this and stands on an ancient ceremonial site going back at least three thousand years. Beneath the present cathedral is one of the largest and oldest crypts in France and a sacred well. A Celtic tribe known as the Canutes once honored the sacred spring and built the well. They masterfully constructed the crypt to carry sound throughout its entirety.[28] Sound is vibration and movement, and from sound energy arises.

Many indigenous peoples were seasonally nomadic and came across these special places on their annual journeys. Places of power were located in areas where lines of earth energy gathered. Earth energies refer to places on the Earth where the energies are particularly powerful. Some of these are lines that link sacred spots, in much the same way that acupuncture recognizes invisible channels that connect key areas in the human body.

The ancients, who were highly sensitive to earth energies, had a "mind map" of these energy lines. More knowledgeable community members, shamans, druids, or medicine men and women had an even more exacting map covering greater distances. Later, as humankind gave up the hunter gatherer lifestyle and tended to stay in one place for longer periods of time, they still longed to make special trips to these sacred sites. Sometimes these trips would align with the seasons and migration of animals, as in the old days. This is how and why pilgrimages came about. Pilgrimages benefit people but they also benefit the Earth.

Sometimes, a person or group of people can instill their energy into the land. For example, there is a labyrinth that I have helped to maintain on a hillside near our home. Many people have used the labyrinth and placed their intentions there. For example, I use the labyrinth walk to recite an affirmation and gain support from the four directions. Yesterday when I visited the labyrinth, I found a somber note tied with a pink silken ribbon to stone. The note was a lament from parents mourning the loss of their daughter killed in a rafting accident. They treated the labyrinth as a sacred place to honor their daughter's passing. This labyrinth has a significant amount of "infused" energy, intentions, and prayers of the people who have walked its curving stone pathway. The people have filled this pathway with prayers.

As I Stood in Chartres Cathedral, I was stunned by the strength of the energy emanating from the ground. I went away from the labyrinth to the side hall. In a few minutes, the tingling vibration stopped. I walked down the long hall, and as I neared the back of the cathedral, a strange figure caught my attention. At the top of the steps, a shimmering, bright white being appeared. While human-like in size and shape, the being was ephemeral and seemed to be composed only of light. The strange figure communicated to me telepathically, with these words: "*You will be a spiritual teacher one day, and you will help many.*" Then the ethereal figure disappeared as quickly as it had appeared.

I stood in silence, trying to understand the meaning of the being's words before walking up the steps to the chapel hallway. At the top of the steps, I looked for the being but there was no one there and no way that anyone could have walked or run to get out of sight so quickly. I knew at a deep soul-level that this was a significant calling to service, but I did not understand why I should or how I could help. My present work was as an environmentalist, saving wildlife and their habitats. My role was protecting and restoring species from the destructive acts of human hands. Being an environmentalist was my life's mission and a very needed one in the 1980s.

The creation of literally thousands of environmental organizations seeking to "save" species and habitats had not yet occurred. Working as a full-time environmentalist was still relatively rare.

The "Conservation Movement" had been going on mainly in the background for 200 years. Eventually, the movement picked up steam, with John Muir and President Theodore Roosevelt conserving thousands of acres of land. Meanwhile, Sir James Ranald Martin led the charge to protect forests and stop extensive tree cutting in Britain, and the Royal Society for the Protection of Birds was created in 1889. Sir Dietrich Brandis, a German forester, developed legislation to protect trees. Yet, it was not until the late 1960s and 1970s that working and getting paid to protect the environment would be an actual occupation for more than a handful of people. The first Earth day in 1970 was one of the first environmental campaigns to reach the general public broadly. By the 1980s, environmental organizations began to form, with a jump from 600 to over 4,000 in just four years, from 1984 to 1988.[29]

At the time of this "celestial" apparition, there was still so much more work to do to protect the environment. Yet, unknown to me, then, the world was entering a time of great changes. The animals and trees would struggle for survival in this new world, and many millions of trees would die and many thousands of animal species would disappear. Environmental changes would be so rapid and dramatic that they would take a decade or less.

It would be many years before I came to understand that our destruction of the planet hurt humankind's heart and soul. Too often, we do not value what we have until it is gone. As the birds, animals, and trees disappeared, many people would begin to regret their loss. Some would even seek out ways to reconnect with the Earth, the trees, and the animals. Perhaps as well, some epigenetic or past-life memory of once being a witch or having a family member treated as a witch shadowed my soul warning me away from spiritual work of any kind. Yet this was all distant from my conscious mind at the time

Thus, the unearthly visitation stayed in the back of my mind for many years. Yet it was well over a decade before its words came into being. Following the spiritual calling has largely been an inward alchemy spurred on by mystical encounters with the trees, the animals, the land, and profoundly gifted human teachers.

Receiving these prophecies about my life path and the vision of a cosmic event on a site sacred to the Celts in the house of the divine feminine was fitting. Today, as the majority of people on the planet come to terms with the reality of global warming, we must also recognize the need to bring feminine forces to bear, as well as the empowerment of women, to help solve this global crisis. As I left the cathedral later that morning and stepped out onto the street into the sun, the street name "Rue du Cheval Blanc," the Street of the White Horse, shone on the sign before my eyes.

13. Spiral Dance

At the still point of the turning world. Neither flesh nor fleshless; Neither from nor towards; at the still point, there the dance is, But neither arrest nor movement. And do not call it fixity, Where past and future are gathered—neither movement from nor towards, Neither ascent nor decline. Except for the point, the still point, There would be no dance, and there is only the dance.[1]

T.S. Eliot

One sunny June morning, a decade after my flight over the Laguna Madre, I stood on a high ridge overlooking a small valley covered with golden grasses in Marin county, California. The hillsides were dotted with groves of madrone, coastal redwoods, and California bay trees. The gentle wind, filled with the Pacific Ocean's salt spray, felt cool on my face. A small flock of western bluebirds flitted above the golden grasses' tips and landed on the branches of a tree. Their bright blue plumage was even more radiant against the red, flesh-like bark of the madrone.

As I watched them, a shimmer of silver above caught my eye. Looking overhead, I saw an osprey flying slowly towards me. Its long, slightly curved wings and a streamlined body appeared like a cross suspended in the heavens. Held in its strong talons was a newly caught fish. Silver flashes gleamed from the fish as the sun's rays landed on its shimmering, still-wet scales. Still alive, the fish was moving its tail back and forth as if it were still swimming. Held aloft with its valiant head forward, its open eyes stared into the great innocence of the sky.

For an instant, I felt pity for that fish. That initial emotion soon turned to one of wonder. The osprey did not carry the fish off in a straight line to its nest on the far hillside. Instead, it flew in slow-moving circles above me, creating an invisible spiral that expanded as it went higher and higher. The osprey was giving that fish a majestic view over the green and gold hillsides, over a doe and her two spotted fawns, over a long-eared hare standing up on jumpy hind legs, and over a wonder-struck woman and her dog.

At that moment, the wisdom and grandeur of that act revealed a timeless secret. It was if the Holy Grail itself had been held aloft pouring its secrets into my mind. My consciousness expanded, and I was lifted beyond every day to a vision of a far grander reality. As the bonds to my body released, my spirit soared with that osprey and that fish. Like that fish, I saw the body of the Earth revealed as if for the first time. This sight, a marvel so beyond the fish's watery dreams and so beyond my own realm of experience, lifted us both up into an unimaginable vision. On that morning, the Creator of all things transformed me with that spiral flight.

The overwhelming depth and breadth of knowledge about the importance of the spiral's form and meaning was absorbed by my soul. After the osprey left, everything I observed around me held within its core design the spiral. Each blade of grass, each tree branch, and even the trees' very trunks rose from the ground as joyous spirals. The gentle wind and the flight paths of the bluebirds also now hinted at spiral motion within and without. I could feel the spiral's dynamic motion moving like a gentle but powerful force upon the Earth filling me and all creation.

Why is it that the spiral is one of the most powerful, persistent, and beautiful Nature patterns? From luminous spiral galaxies ablaze in color to the tiny spirals at the tips of our fingers, spirals are everywhere. The spiral is both a form and a movement, found throughout every inch of the Universe. When the spirit moves in life-affirming, life-generating ways, it moves as a centripetal spiral, internally towards the center. The spiral is nothing less than Creation itself.

Human blood, which is 92 percent water, flows in spirals through our arteries and veins. This spiral movement is also responsible for the movement of blood throughout the body, not just the heart.[2] According to Viktor Schauberger, a Bavarian forester, water, being akin to the blood of the Earth, needs to flow in spirals to generate and maintain life force.[3] This water, which he called "Living Water," is one of Earth's most important life-giving substances.

Schauberger discerned that water is also better able to support aquatic life when it moves in a hyperbolic spiral towards the center. When the spiral moves inwards towards the center, it causes growth, unfolding, evolution, and life. When the spiral moves in the opposite direction, outward, it causes dis-assimilation, decay, chaos, and death. Both forces exert power in our world: one exerts implosive power, the other explosive power. Today, humankind has unwittingly harnessed the destructive explosive force to a much greater degree than its equal and necessary balancing force of implosive power.

For example, the combustible engine, still similar to its original design from 1798, uses "explosive" energy as trillions upon trillions of noisy, mini explosions take place all around the world in cars, ships, planes, leaf blowers, and more. These destructive forces far outnumber the uses of the more life-generating implosive power in our industrial societies.[4] According to Schoenberg, "*When our technology only uses the decomposing motion, it becomes a dead technology, a destructive one, dangerously affecting all of Nature.*"[5]

Meanwhile, the life-generating and organizing aspects of the implosive or center-seeking spirals abound and surround us in Nature. Plants grow in slow-moving spirals toward the heavens, and flowers and snail shells flow in a spiral growth pattern. Even our bones grow in spirals to more effectively support our weight and mass.[6] While spirals often go unnoticed by our conscious mind, we are subconsciously aware of the naturally occurring spiral shapes around us in Nature. This is one reason why spending time in Nature can help establish equilibrium within the body and mind.

The blueprint for all life on Earth exists in a spiral form: Deoxyribonucleic acid or DNA. DNA's centripetal spiral is the movement and shape of Creation and life. This spiral is the manifestation of the process of becoming, just as life arises from movement arising in the womb of the primordial Ether. The spiral is life itself. Ancient peoples understood the power and prevalence of the spiral. They captured its essence in their designs and in their rituals and spiritual gatherings. Its shape is found in thousands of places replicated by human hands in Africa, New Zealand, and in Tibet. Those who understood the secrets of Nature wished to depict the dance of the Universe for future generations.

Human movement follows a spiral pattern. Great artists understood the inherent beauty and symmetry in the spiral and depicted the human body in a subtle, helix position. This can be seen in the works of Michelangelo and Auguste Rodin.[7] When we use spiral movements in sports or dance, our actions are more powerful and graceful. Aikido is an excellent example of an art of movement where students are taught to imitate the spiral form and to take advantage of the spiral energy within their very being.

The spiral dance takes place within our own psyche as well. Creation stories have cyclical patterns, just as human psychological and spiritual growth follows a spiral path. Human nature is a flowing and ever-enlarging adaptation to the outside world and internal learning. As we transcend, psychologically, and spiritually, we build upon previous value systems much in the way that a tree grows around its central core. We are forever expanding outward and upward, the inner rings of our being serving as a foundation for growth. Like a tree, once people grow in their psychological development, this growth becomes the foundation for their future evolution. As we evolve, we become the heroes and heroines of our own lives.

Just as our psychological growth comes about through a spiral evolution, so does our spiritual understanding. When we move into a sacred place within our being, our spirit can also rise like a spiraling snake. Kundalini energy, divine feminine energy and subtle force within the body rises from the base of the spine and climbs spiral-like up through all the chakras before it bursts through the chakra at the top of the head. Ultimately, the entire human race, and even the Earth, go through a spiral evolution of sorts. Sometimes, this can happen slowly over time, yet great changes can occur instantly, shaking one to the core of their being.

~

As the Milky Way moves through space towards Andromeda Galaxy, 2.5 million light-years away, our sun orbits the galactic center, and the Earth orbits the sun as our entire solar system moves in a helical spiral motion through space.[8] We hurtle through space at 1,700 kilometers per hour, as the Earth rotates on its axis.[9] Space is not uniform but contains different qualities of energy and subtle vibrations throughout. As our solar system moves and enters these new regions, everything within the solar system, including the Earth, is bathed in these new cosmic emanations. Many ancient peoples like the Hopi believe that as we enter new energetic vibrations in the Universe, this triggers the evolution or devolution of the human consciousness and body. Some say that time is now.

The Hindus believe that humankind evolves through great ages or Yugas. Just as humankind's consciousness spirals upwards from its initial focus on gross matter, or the lowest consciousness, to its pinnacle, a focus on the nature of subtle energies, it then will return and devolve back to slip once again into ignorance and darkness.[10] They believe that this continual spiraling up and spiraling down is in direct relation to the location of our planet to the central sun of the Universe. This central sun is envisaged as the mother star of our galaxy, much in the same way that a nucleus in a cell is its focal point. This central sun is called Brahma by the Hindus, and it is Brahma that regulates and evolves dharma, or mental virtue and understanding. When our sun and solar system are close to the mother star, all living things on the Earth are elevated in their understanding. Humankind will become highly developed, and all things will be easily comprehended, even the deep mysteries of subtle energies and the soul. Yet, humankind cannot "rest on its laurels", evolution can be and has been stopped and ended for many species.

By understanding Natural forces like the spiral, and working them, we shift the outcome in our favor. Ultimately, the spiral teaches us that our path is the same as the path of all living beings, a path of evolution, unfolding, and transformation. When we enter the sacred realm of the life-generating spiral, we find the Holy Grail: we discover the intimate secrets of life itself.

IV. EARTH

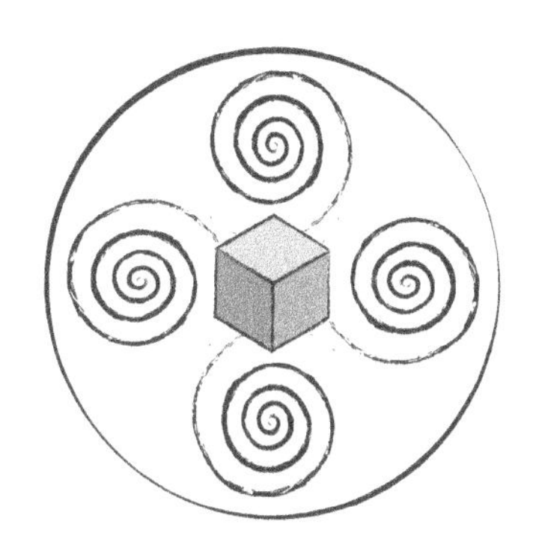

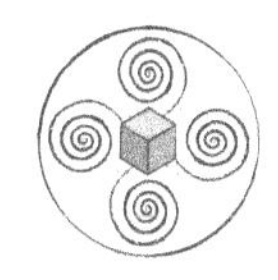

14. Aluna

Some trees are like your mother or father.[1]

THE KOGI

In 2014, I spent an unforgettable afternoon and evening with two indigenous leaders from Columbia. I had been invited, along with 11 others, to listen to their story and learn about what is happening to the Earth. The Kogi are the last surviving society of the ancient pre-Columbian peoples that has continued their cultural, political, and spiritual beliefs and practices without interference from outside societies.

The Kogi live in the heart of the wilderness, presently known as Sierra Nevada de Santa Maria, the highest coastal mountain range on Earth. It is a veritable planet in miniature since it contains almost every ecosystem possible. The Kogi's remote and densely forested home has kept out invading conquerors, missionaries, and gold-seeking looters for thousands of years. Their location has protected them from destruction and assimilation by other cultures, and it has become their sanctuary. Theirs is one of the only "untouched" cultures in the world.

The Kogi believe that their longevity is due to listening and learning from the Earth. They believe that the Earth is a living being and part of a larger cosmic consciousness. They refer to this consciousness as "Aluna" and say that one can converse with and learn from her. The Kogi have a profound understanding of ecological principles, human nature, and even the stars. Their innate knowledge of the Universe, and their application of this knowledge to their own lives, is reminiscent of other ancient societies that honor a living Earth and intelligent cosmos.

For many years, the isolated Kogi noticed the accelerating degradation of the planet. They learned from Aluna that modern industrial societies are causing massive destruction of the planet. The Kogi refer to members of these "industrialized" societies as "younger brothers." So named, because people from these societies are ignorant of the world's ways. Even though the Kogi prefer to live away from these societies and not have contact with them, they are so alarmed about global warming and other environmental ills; they decided to reach out to us, the "younger brothers."

While the Kogi had never operated machinery or even seen a camera, they understood, through divination, that this was a powerful way to reach people in industrialized societies around the world. Thus, they asked British filmmaker, Alan Ereira, to help them make a film to send their message and warning to industrialized man. The documentary From the Heart of the World was released in 1990.[2] Once the movie was completed, the Kogi withdrew again from all contact with modern society.

Yet, they felt called to reemerge when they felt that "younger brother" was not heeding their message. Their second film, released in 2012, is called Aluna. In Aluna, they tell us that the obvious signs of droughts, floods, pollution, and snowmelt are, to this tribe, powerful indications of a dying planet.[3] The Kogi feel that we must understand that the Earth is alive, that sacred places on the Earth are connected, and that these places are being destroyed. They are driven to help us see and understand what they plainly know. In Aluna, the Kogi show how places on the planet are inextricably connected to others. Just as our bodies need all of our parts intact, so does the Earth.

They illustrate, for example, how the damage caused at the coast, at the mouths of rivers, by construction and mining directly results in destruction at the top of the mountain. Once beautifully capped in white snow, their mountains are now brown and bare, while mountain lakes are dry, and the trees and vegetation are dying. They know that if you dry up the estuaries, you affect the upstream source of the river. These understandings were confirmed in the movie by Alex Rogers, a specialist in ecosystem restoration, professor of zoology, and world leader in marine biology at Cambridge University in England. According to Rogers, "*Their view that all these activities are having an impact at a larger scale are quite right.*"[4]

~

On a June morning, I arrived at the private donor's home in Northern California to meet the two Kogi leaders. The large home was surrounded by a lush green lawn circled by pathways with graceful trellises. There was a buffet lunch prepared, and our group sat together with our plates on our laps in a circle under one of the trellises looking forward to meeting the Kogi leaders. After lunch, the leaders arrived, and I was introduced. One of the Kogi leaders was referred to as the "Mama." Along with the Kogi Mama, the Kogi choose a "political leader" to jointly guide their tribe. Therefore, I was meeting with both leaders, the present spiritual and political chiefs.

The Mama is the spiritual leader of the community. Mamas undergo strict training to assume this priestly role. Unlike traditional religious leaders who read from books, attend to the rules of their order, and ascend a political ladder, Kogi priests are expected to communicate directly with Aluna, the Mother Goddess. Kogi Mamas learn to connect with cosmic consciousness to keep the world in balance. Kogi believe that communicating with Aluna is one of the reasons for our existence and our critical role on the planet. Children selected as future Mamas may begin their "training" as early as the age of two. They are kept inside a darkened home for the first nine years of their lives. Elder Mamas and the child's mother care for, feed, train, and teach the child to attune to Aluna. Through living in quiet solitude, learning to listen, and by protecting their senses from bright lights and spicy foods, they come to hear Aluna through finely attuned sensory perception.

The Kogi wear simple clothing that they weave themselves. I was struck at how purely white their simple cotton robes appeared and how similar in appearance they are to the robes worn by the Pictish Druids.[5] I had brought the Kogi some simple gifts of unbleached cotton fibers for weaving and wooden beads. I brought them some turkey feathers that I had found during one of my many hikes into the woods. The feathers were in beautiful condition and shone brightly with purple and blue glistening colors in the sunshine.

The Kogi graciously accepted my gifts even as the translator misinterpreted my explanation of the feathers' symbolic meaning. I had tried to explain that the gift of the shiny turkey feathers instead of Eagle feathers was a way of extolling the qualities of the humble "Earth Eagle," despite America's official adoption of the sky eagle as the nation's symbol. The Kogi leaders nodded quietly as they looked deeply into my eyes, seeming to understand what was lost in translation intuitively. They accepted the gifts graciously, and we shook hands.

After everyone was introduced, the Kogi got up and walked away together slowly around the large estate grounds, occasionally stopping to confer quietly between themselves. After an hour or so, they came back and asked for us to follow them. We all stepped under a trellis and walked away from the large property's manicured part to join them at the edge of the property. The Kogi stopped and stood next to a monolithic-looking, six-foot-high stone as if they were presenting it to us. We gathered in a semicircle around the stone and the two Kogi leaders. After a few minutes of silence, they asked us to *"empty our thoughts into our cupped hands."* We did this by focusing our thoughts and intentions and staring into our cupped hands, imagining our thoughts pouring into them. Once we had done this, we were asked to "pour" our thoughts into a large, dark-green leaf that they had folded into a cylinder for this very purpose.

As our group took turns carefully "depositing" our thoughts into the carefully folded leaf, the Kogi leaders watched each one of us intently. After everyone was done, they picked up the leaf and examined it quietly together. They said a few words to each other as we waited and then led us back towards the house for their "talk." Since they speak Taironian, there were two translation levels needed, one from Taironian to Spanish and then another from Spanish to English. Thus, their talk took several hours to convey their knowledge and their wishes. While some of their words and intent may have been lost in translation, below is some of what they shared with us that day.

They asked us to open our eyes to hear Mama's story about how things really are. They explained that Aluna is the Great Mother, conceived by water and earth. That she is inside all of Nature and that she is fertility and intelligence. They said that they are the Elder Brothers first created to take care of the Earth and that much later, we, the "Younger Brothers," were created. They said that we Younger Brothers (people of industrialized societies) were causing the Earth to lose her strength because of greed. The Younger Brothers do not understand that payment must be made for what is taken from the Great Mother. The air we breathe, the water we drink, and the animals we kill need to be honored and acknowledged. They said the Younger Brothers do not appreciate these things as they should be.

The Kogi warned that we, Younger Brothers, had forgotten the importance of the waters. They explained that they learn and even foretell future events by consulting the water and that the water is alive. They warned us that the water was being harmed, that humans must have water to live, and that humans, animals, and trees dry up and die without it. They said that the destruction of the estuaries and wetlands in the low regions destroys the waters upstream, where the rivers are born. They also warned of many new diseases that will arise because of the harm to the water and the planet and that these diseases will not be curable.

They admonished us not to cut down certain trees. They said that "*some trees are like your mother or father*." They said that they did not cut down the Banyan tree in their forest. They said we could not cut down one forest here and plant another there, thinking that no damage is done. By cutting that first forest, we damaged a critical organ of the Earth or destroyed a sacred tree, which is the father or mother of the species. Or perhaps we destroyed an Esuana. An Esuana is a primal, energetic thread that holds together all of Nature. They asked, "How do you know you are not cutting down the grandmother or grandfather tree that all the other trees of that kind depend upon?"

They said that certain places in the forest golden images of the animals were hung from trees by their ancestors or buried there by an Esuana to protect the spiritual home of a species. Some of these golden images have been discovered and stolen by younger brother lusting after the gold and not understanding the honoring's spiritual significance. They said it was wrong for their Younger Brothers to take these gold "honorings" away from the animals. These offerings at the "home" of the animal protected the entire species and must be returned.

They explained that certain places on the Earth where energy lines run along with geographic areas, where the metaphysical plane binds strongly with the physical world and holds it together. They told us that they knew where their ancestral spirit paths, lines of energy laid down by their ancestors' thoughts, ran. In the Lost City, an ancient Taironian village, a map stone has been found that indicates how and where lines of intelligent energy connect sacred sites. The placement of Taironian buildings and routes was carefully calculated to reflect larger invisible energetic realities and spirit lines. The maps were discerned by the ancient shamans and still guide the Kogi today.

The Kogi said they have an invisible map of our world and the larger cosmos. That if these lines or the places along these lines are damaged, this can unhinge the many threads of energetic connections held in locations throughout the Universe. These places of energetic connection between the manifest and invisible worlds support the planet and life in the solar system and beyond.

The Kogi said that we could no longer cut down forests and drain swamps, dam rivers and fragment ecosystems with roads, or dig pit mines and drill gas wells with impunity. To do so in one place can damage the entire Earth. Just as if you cut off a person's limb or remove an organ from their body, these cannot be replaced; the same is true if one damages certain places on the Earth. The Kogi said we are close to the "dying of the world." They explained that many things have died and will die because of Younger Brother. That Younger Brother does not now understand the extent of the damage, nor the coming chaos, will ensue. They said to us, *"If you knew how she [the Earth] could feel, you would stop."*

Yet, they also encouraged us and said that the world could "go on" if we act well and stop our destructive activities. They encouraged us to grow in our spiritual understanding, learn to be grateful to Mother Earth, accept the responsibility for our actions, and develop our consciousness and understand the great power we have to either save or destroy our world.

Mamas believe that their role is to support and guide their community and support the Earth and all creation. Kogi believe that it is important to live in balance and harmony with Nature and they incorporate this belief into everything that they do. For example, there are some trees that are considered sacred and never cut. Protecting waterways and keeping the water clean and pure is another mandate.

Kogi's also work with the ineffable. They insist that focused human intentions for balance and wisdom are required to stop the world from descending into chaos. They believe that it is the "*great mother's thoughts*" that keep the world "*together in harmony*." Aluna's thoughts may make the Earth manifest, but that it is humankind's role to support and join in Aluna's thoughts that keeps the Earth vibrant and alive

They believe that even people from industrialized societies, like the younger brother, can learn to think and act in a way to support a living Earth and that it is humankind's obligation to do so. This must involve a shift in western people's consciousness, not just a change in how we conduct outward activities. They explained that the material world we refer to as the "real" world does not stand alone but is embedded in a larger cosmic world of pure thought.

According to the Kogi, "*Aluna needs the human mind to participate harmoniously in the world or chaos will be unleashed.*" All human beings have the capacity to enter into a higher state of consciousness and access the cosmic mind. They told us that as human beings, we have an obligation to protect these important places and lay down positive energy lines with our thoughts and honorings. They reminded us that our ancestors understood the importance of thoughts and laying down positive energies and lines to give life to the Earth. They told us that these positive thoughts of mother Earth and the animals could manifest in the world, bringing life, balance, and beauty.

They said that if we understand the planet as a living entity and Nature as intelligent, we will begin to ask, "*What does the Earth want? What does the river want? What does the tree want?*" By doing so, we will be able to live in harmony and balance with the natural world and lead more fulfilling lives.

Then it would not just be laws and regulations that stop us from damaging the environment; it would be our understanding that to dam a river would be like cutting off our mother's lifeblood, to destroy an estuary would be like removing one of her arms, or that to drill for oil along one of her sacred lines would unhinge the entire world. They stated that we must understand the depths and reach of the principle of interconnectedness.

As they ended their talk, each of us sat in profound silence.

As I drove towards home that evening, I knew that their words validated many of my lifelong feelings about and experiences with Nature. Their words resonated deep within my consciousness, ringing deep into an ancestral past. Yet, I wondered: was it truly possible for westernized people to change in time?

Many thousands of years ago, our ancestors lived like the Kogi, close to the land, following the seasons, watching the bright stars in the sky, and learning from the trees, the animals, and the Earth. Their wisdom is still here, buried deep in our DNA and circling the planet in intelligent clouds of thought.

We are fortunate that the Kogi, the wisdom keepers of Aluna and the Earth, can share their wisdom with us. My only hope is that their wisdom does not fall on closed ears and minds. Approaching the animals and the earth nakedly, with an innocence and an openness to what we could learn and be, is a Calling we must attend to.

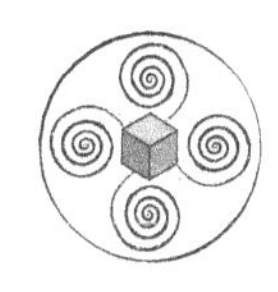

15. Plants: Earth Healers

The forest is a peculiar organism of unlimited kindness and benevolence that makes no demands for its sustenance and extends generously the products of its life activity; it affords protection to all beings, offering shade even to the ax-man who destroys it.[1]

Gautama Buddha

As she reached out her small hand, I said quietly, "*Look into its face before you pluck the flower and thank it.*" The quizzical expression on Amy's face led me to explain further: "*Everything is alive and filled with life source. By acknowledging the plant and thanking it for its beauty, you show your appreciation for the good things that Nature freely offers. It's like thanking a friend for giving you a lovely gift.*" It was a beautiful April morning as I led a youth quest nature program for girls aged nine to fourteen in a state park in Northern California.

Indigenous people provided their children with education about Nature, throughout their days. Children learned how to find food and how to build shelter, which plants and herbs were suitable for healing, and in doing so, they came to know and understand everything that walks, flies, or grows. This, Naturalist intelligence, a term coined by Howard Gardener,[2] is the ability to recognize everyday natural objects and living things, like insects, branches, plants, rocks, and animals. It is in this area of specific intelligence that humankind is falling behind. In large part, this is simply because we lead lives so distanced from nature and spend so little time outdoors. There is little being offered to children today to make up for this astonishing lack of awareness about the very place that we live in. As the planet becomes more industrialized and computerized—we lead lives less connected to the land.

Beyond the necessary scientific and practical knowledge that comes from this knowledge, there are a multitude of advantages of spending time in and learning directly from nature. Time spent in natural settings has been proven to heal humans physically and psychically. While this is something that people have intuitively known, numerous recent scientific studies have proven the benefits of Nature on our physical and mental health and spiritual development. In Growing Up, Naturally, researchers Eric Windhorst and Allison Williams found that university students who experienced Nature deeply as children tended to be more positive and have higher life satisfaction and greater vitality.[3] Their study concluded that a connection to Nature supported a child's emotional and physical health even into adulthood.

A larger study carried out by The University of Exeter Medical School reviewed over 10,000 city dwellers' mental health data over an 18-year span. It revealed that people living near parks and open spaces suffered less mental distress.[4] Similarly, Dr. Jolanda Maas of the University Medical Centre in Amsterdam found that living near green spaces significantly lowered mental illnesses such as depression and anxiety, as well as physical ailments such as heart disease, diabetes, asthma, and migraines.[5] The team analyzed medical records of almost 350,000 people.

The Japanese scientist Yoshifumi Miyazaki has been conducting detailed experiments on the restorative aspects of walking in a forest. Not content to just ask people how they feel, he tests hormones and other bodily health indicators. He has found that walking calmly amongst the trees decreases stress hormones like cortisol by 12.4 percent.[6] Just a quiet woodland walk can result in a 1.4 percent decrease in blood pressure, a 5.8 percent decrease in heart rate, and a seven percent increase in sympathetic nerve activity. He also found that natural immune cells, sometimes referred to as NK cells, increase as much as 40 percent after a person spends several days in the woods. This boost in the immune system lasts for a month or more. Along with these physical health benefits, participants in his study reported that they felt less anxious and happier.

All people should have opportunities to learn from nature firsthand. With 90 percent of Americans living in an urban area, there are fewer opportunities to have experiences in Nature. Environmental education provides opportunities for people to forge greater connections to nature and to learn from nature. David Brower, one of the world's greatest environmentalists and mountain climbers, said that some of the most important lessons that he learned in his life were learned from nature. In an interview that I conducted with Brower over 20 years ago, he gave an example of watching a butterfly struggle to escape from its cocoon and how to rush along the process. He "helped" the butterfly by removing the cocoon for it. Yet the over-hasty removal left the butterfly's wings damaged beyond repair and the insect unable to fly. For this, he learned that some things must happen in their own time and way. This was a simple lesson, but it became profoundly meaningful through this direct experience with the natural world.

Like Brower, Nature was my teacher. I could spend hours even as a wee girl of 5 or 6, patiently watching and learning from the animals and the trees. This is how I learned that the exotic-looking red-winged black bird's mate was feathered in quiet browns to stay hidden while on her nest. These direct experiential learning experiences stay with one for a lifetime. They also, no doubt, assisted in the growth of my cognitive and information synthesizing abilities. This is one reason why I began offering "Youth Quests" and "Girls Quests" outdoor education programs. Important in its own right as a rewarding time of personal growth, early adolescence is the phase when young people adopt belief, behavior, and learning patterns that can have lifelong consequences.

Engaging children in natural settings and providing them with hands-on learning experiences accelerates creative imagination, self-confidence, and active play and exploration. A large part of Girls Quest outings is simply allowing the children to spend time in nature and experience it first hand, through exploration and observation. According to a comprehensive 40 school study by the State Education and Environment Roundtable "SEER", young people learn more effectively within an "environment-based context." Schools reported that their students engaged in outdoor and environmental education saw improvements across many academic topics, including science, math, critical thinking skills, and more.[7]

~

This April morning, I was standing in the middle of a circle of 26 girls from eight of the local schools, the aroma of wildflowers, blooming in shades of yellow, white, pink, blue, fuchsia, and purple encircling us. Fourteen-year-old Amy, one of the participants, had just come across a "baby blue eyes." Formerly known as Nemophila menziesii, a lovely, small, six-petal blue flower. Inspired by its enticing loveliness and subtle color, it was so lovely that she felt impelled to pick it. When leading youth quests, I usually ask youth not to pick flowers but to leave them to feed the bees and butterflies. Today, however, there were so many of these flowers forming a carpet of blue around us that just one flower could make this a teaching moment.

One of the things I love about flowers and plants is how they interact with other living things. Like humans, plants seek to pass along their DNA to future generations, yet while wind and water can move pollen from one plant to another, these are not always reliable methods. Thus, 130 million years ago or so, plants created flowers to entice insects, birds, and even mammals to help the process along.[8]

Bees, hummingbirds, and even small mice gather the sweet nectar from a flower and then invariably pick up pollen, helping the next generation's proliferation. It's an ingenious and lovely way to help to ensure the continuity of a species. Best of all, it is based on "mutualism." Mutualism is the idea that different species benefit each other.[9] The plant provides nectar and pollen that can be used as food by insects and animals, and in exchange, the insects and animals assure that the plants pass along their genetic material.

Of course, this is not the first, nor the most important, mutually beneficial activity that plants engage in. The plants first created an atmosphere that allowed complex life forms like mammals and humans to exist.[10] Not only is the oxygen in the atmosphere critical for human life, but oxygen is also the most prevalent element found in our bodies.[11] With every breath we take, we owe our existence to the plants, from the tiniest photosynthetic algae in the sea (which create more than half of the oxygen on the planet) to the giant redwood trees.

Plants achieved this singular "victory" in making Earth "livable" for advanced life forms by tapping into an unlimited energy source, the sun. The feat of turning solar photons into life energy and chemicals in a billionth of a second is nothing less than extraordinary. Thus, we may enjoy the beauty and delightful scent of a flower, but there is great intelligence behind how that flower came to be and why it exists.

Solar panels do what plants do by capturing sunlight and turning it into useful energy. Surprisingly, while solar is clearly one of the most sustainable energy sources on the planet, it is still eons behind living plants' efficiency. Today, most solar panels capture just 20 to 30 percent of the photons that reach them from sunlight to be converted to energy, while plants capture 95 percent.[12] For many years, scientists could not explain the higher return rate that plants achieve. It was a perplexing mystery because the probability of such a fantastic efficiency rate was almost zero based on purely linear mathematical calculations.

However, an international team of scientists had a breakthrough in 2014 when they claimed that plants use quantum coherence and superposition to obtain a solar absorption rate beyond linear possibilities. According to research published in Nature, all plants, from single-cell algae to tall redwood trees, transform sunlight into energy stored chemically as carbohydrates.[13] The quantum key to doing that lies in a phenomenon known to physicists as quantum coherence:

Quantum coherence describes how more than one molecule interacts with the same energy from one incoming photon *simultaneously*. In essence, rather than the energy from a particular photon choosing one route to pass through the photosynthetic system, it travels through multiple channels simultaneously, allowing it to pick the quickest route. It is as if you could drive to work via three different routes *simultaneously*, losing no time or energy to traffic delays on any of the given routes.[14]

The solar energy transfer is extremely efficient, with almost every photon being captured by the plant's cells. The plant cells use molecular vibrations, or "beats," at a quantum level to attract individual particles of sunlight (photons). These vibrations guide the individual photons to a specific destination on the plant's leaves where they can be "captured" by photosynthetic cells and turned into energy. The plants are "singing" for the light!

~

We may see ourselves as a far superior species to the plants, yet the reality is that they have made our planet a safe and nurturing place for humankind to survive in; yet, the plants would be fine without us. Plants can teach us the benefits of engaging in "mutualistic" ways of relating to other life forms. Perhaps we can begin by looking into their flower faces and thanking them for all they do to benefit the Earth and us.

Plants could also be our greatest allies in the first slowing and then reversing climate change impacts. How do we know that plants, from the tiny phytoplankton to the largest redwood trees, can help us heal our damaged atmosphere and restore a healthy climate? Because they have already done so, more than once. Of course, plants initially created the oxygenated atmosphere allowing more complex and diverse life forms like mammals and humans. Thereafter, they have modified the planet's atmosphere and weather to be optimal to support life as we know it.

Creating and maintaining an oxygen-rich atmosphere is just one way that plants have made the planet a hospitable environment for life to exist. There are many other ways that trees have positively shaped our world. Vascular plants terra-formed the Earth's surface to make it more conducive to life. Plants created marshes and swamps and those fertile places between the waters and the land. Between the Cambrian and Pennsylvanian epochs, vascular plants were hard at work, turning rocks and hard substrate into mud and soil.[15] Working with powerful root systems and chemicals that could erode rock, they alfso developed sand bars and banks along rivers. This led to stable riverbanks and even braided rivers with islands of vegetation along flood plains. These became the birthplaces for all types of aquatic and semi-aquatic animals and insects and are still critical to many specie's survival. Plants play an essential nursemaid role, providing abundant food for humans, animals, birds, insects, and reptiles in the form of fruits, nuts, acorns, flowers, foliage, and nectar.

Trees also play a critical role in cleaning water and protecting aquatic life. They provide shade for streams and other water bodies. This keeps the water's temperature cooler, reduces evaporation, and increases the water's ability to support aquatic life.[16] Along the streams and rivers, referred to as riparian buffer zones, trees and other plants stop contaminants from entering the water and act as natural filters.[17] Plants have also been used to clean wastewater at a fraction of the cost of a traditional sewage treatment plant. Referred to as "living machines" natural waste treatment operations use plants, the sun's rays, and even fish and snails to turn filthy, polluted water into clean, potable water for drinking.[18] The plants have turned the Earth into a habitable home.

It was the plant Azolla which helped moderate the Earth's climate. Fifty-five million years ago, the Earth was dangerously overheated by greenhouse gases. Reaching what was known as the Paleocene-Eocene Thermal Maximum, greenhouse gases were admitted into the atmosphere and drastically changed the atmospheric conditions and the Earth. Massive extinctions occurred in the oceans and on land.

Then, Azolla, a simple water-friendly fern in the family Salviniaceae, came to the rescue. Azolla thrived in a warmer environment and took over places where other plants failed. Azolla has superpowers when it comes to removing carbon and nitrogen from the atmosphere. As the plant's population exploded on the planet, it accounted for an 80 percent drop in CO2 in the atmosphere! Under the right conditions, meaning sunlight and phosphorus availability and other needed nutrients, Azolla can double its size in two to three days. Thus, along with the Earth's plates and ocean currents' movements, the plant played an important role in cooling the planet and reversing the world's greenhouse state 48 million years ago.[19]

Could this amazing plant species, along with many others, help us mitigate global warming today? While we now face a more perilous situation, since the global changes were occurring ten times faster than 55 million years ago, the answer is yes! Yes, as long as we stop cutting down the forests and polluting the oceans. Yes, as long as we stop emitting carbon dioxide and methane. Yes, as long as we start saving old-growth trees and forests and planting new trees and forests. Yes, as long as we stop tampering with geoengineering programs that dim the light from the sun and rain down heavy metals like aluminum, which ruins the soil and impede plants' ability to absorb phosphorous and other nutrients. Yes, if we can help the plants to do what they do best naturally: change the climate for the better.

Plants first created the oxygenated atmosphere on Earth, and they also terraformed the planet. They did this through photosynthesis and by causing chemical weathering of silicate rocks with symbiotic fungus.[20] They also regulate the weather by way of transpiration and photosynthesis. The tropical forests are key drivers of this important process by creating atmospheric rivers of water.[21]

The point that "plants equaled water" hit home when I worked in Texas as the Audubon Society's Executive Director. A local Texas billionaire, David Bamberger, bought one of South Central Texas's largest ranches. Previous ranch owners had allowed cattle to overgraze and damage the land, leaving the ground dry, with few if any native plants left. Also, the intermittent streams that had once existed on the property were all dried up. Since money was not a limiting factor, and Bamberger wanted to bring the ranch back to life, he hired habitat management consultants to plant tens of thousands of native plants on the property. They focused on planting grasses in the recharge zone of an important aquifer. The grass's roots stopped rainfall from running off the land and helped water to flow down to the aquifer. Within just two and a half years, not only had many species of birds and animals returned to the land because of the plantings, but water appeared once again in the dried-up stream beds and remained.

Plants conduct evapotranspiration, which means that they not only release oxygen but put moisture and particulates into the air. This moisture and particulates can create atmospheric changes. In the case of the Amazon rain forest, where millions of trees draw up water from their roots and releasing it into the air, an enormous "flying river" is created in the sky.[22] This sky-high "river," which can even be seen from space, eventually drops its water as rain onto the Amazon basin, and this nutrient-dense water flows into the ocean. Once there, the "soup" of tree water and sediment supports tiny and unusual organisms called diatoms.

A diatom is a tiny, single-celled creature that lives in a glass house; its exoskeleton is made of clear silica, allowing one to see its internal "organs." This unique part-plant, part-animal algae, supplies the Earth with 40 percent of carbon sequestration and 20 percent of the world's oxygen.[23] The diatom is not the world's only chemical magician. In the oceans, tiny phytoplankton produces organic sulfur compounds as a chemical defense against damaging UV radiation.[24] They will start producing and excreting these compounds (such as dimethylsulfide, or DMS) within a matter of days of being triggered. DMS then produces clouds that can block excess UV radiation.

The Intergovernmental Panel on Climate Change has identified natural aerosols emitted by plants as potentially one of our most important tools for combating global warming.[25] Forests are the sources of up to 65 percent of all clouds and 45 percent of all rainfall.[26] Plants and forests should be at the top of our list of climate-healing heroes, and we should be thoughtfully supporting their survival and increasing their numbers. Terrestrial plants protect themselves and the planet from excessive UV radiation. Pine forest and boreal forest give off large amounts of certain chemical compounds that can significantly cool the climate by reflecting the sun's rays back into space.[27] Thus, North America, Canada, Russia, and Europe landmasses that are home to these forests, could safeguard existing trees and plant additional ones in the region to cool the Earth.

Yet, plants are not immune from rapid and significant climate changes that can "shut down" their epic abilities. This fact was hit home in 2015 and 2016 when the largest single increase in carbon dioxide in 2,000 years occurred, releasing 2.5 more giga-tons of carbon into the Earth's atmosphere.[28] The satellite OCO-2 was able to pinpoint the causes of this significant 50-percent increase. The driving forces were an El Niño year that led to the most severe drought in the Amazon rain forest in 30 years and unusually high temperatures that stressed vegetation and reduced their ability to conduct photosynthesis.

This drought had a net effect of making the plants less able to create oxygen and unable to absorb carbon dioxide. Because of this, Asia experienced unusually hot temperatures and extensive peat and forest fires. Even though rainfall was at normal levels in tropical Africa, the extremely high temperatures felled plants and made them decompose faster. Scientists working to understand the data coming back from OCO-2 now fully understand that plants are major players in addressing global warming.

The important role of plants in addressing climate change is but one of the many things that children and adults can learn by gaining a deep understanding of nature.

Occasionally I hear from some of my former youth quest participants who share that they still remember the things that they learned about the Earth and the animals. A few, now parents themselves, tell of how they take their children by the hand to show them the flowers, birds and trees and see the wonder in their children's eyes.

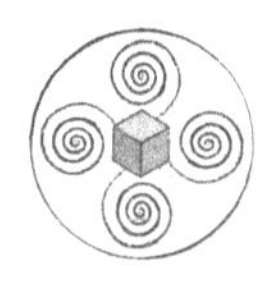

16. Thinking Like a Wild Woman

A nation is not conquered until the hearts of its women are on the ground.[1]

CHEYENNE SAYING

I no longer fly single engine planes. Although I miss the physical experience of flight as a pilot, I encourage and help others to "fly"; not in a plane in the sky, but in their thoughts. I guide people to discover and follow their path with a heart and connect intimately with the sacred through Nature. Soaring on that inner journey, we can find a place within that reflects our authentic selves. When we open our eyes to the amazing life forms that surround us from this place, we can better understand the world around us. Freed from the confines of socio-political forces and limiting beliefs, we can launch ourselves anew into the waiting world. There is no time to waste!

This is why I also lead efforts to transform our old and outdated ways of "interacting with and "managing" wild lands, wildlife and Nature. I advise and conduct habitat restoration and species protection. I educate landowners, and agencies on what the land, the trees and the animals need to survive now that the world is rapidly changing. In 2018, I launched a forest conservation program to protect and restore forest ecosystems and the species that depend on them. Climate change poses new risks to ecosystem health, demanding innovative and proactive approaches to conservation. Yet, many existing laws, policies and practices pertaining to forest and wildlife management inadequately address the stunning changes occurring regionally and globally. Many of today's challenges, like the sixth mass extinction of species, have never been experienced by previous generations.

Direct causes of decline are misunderstood or misinterpreted, leading to inappropriate and ineffective management. The survival of our ecosystems requires a bold approach to quickly and correctly identify environmental changes and their direct causes, and to apply solutions responsive to today's threats to biodiversity. The forest work that I do involves the following four strands: 1. evaluate and assess the health of forest ecosystems (soil, plants and trees, and wildlife), 2. Develop habitat and species Management Plans for Private and Public Lands to support forest health and biodiversity, 3. Educate the Public on the state of our forests and wildlife and introduce them to innovative practices they can use in their own backyards, and 4. advise and prepare new forest and wildlife policies needed for a climate-changed earth.

Today, many women are doing similar work; some are spiritual teachers, some are healers, and some are leading movements and organizations positively transforming our world. An example of women's unique leadership in the environmental field is the non-profit organization Women's Earth Alliance, also known as WEA.[2] WEA supports women in Africa, India, and the United States with training, mini-grants, and advocacy support. For example, WEA launched a project in 2016 to bring clean, safe drinking water to the 6,000 residents of a community in Northern Assam, India. In 2014, WEA trained 50 African women leaders from Uganda, Kenya, and Tanzania to learn how to build water and sanitation facilities for their communities. WEA staff members not only provide local women with valuable skills. By training women to construct rainwater harvesting systems and bio-sand water filters, they help them become self-sufficient. Many also then launch their own businesses, developing these technologies for other communities in their region.

I first met Melinda when I was the Executive Director of an international environmental organization and hired her right out of college. She impressed me with her intelligence, dedication, and compassion. I soon had her flying to Alaska to meet with policy leaders to prepare for our International Bering Sea conference and to China to attend an environmental conference for women. Within a few years, Melinda launched her own international NGO, and I joined her advisory board. She soon found an excellent partner in Amira Diamond, and they now work together cooperatively as cofounders.

Melinda says that she founded WEA because she was '*moved to her core while witnessing solutions that women created in the face of great odds.*' According to Melinda, "*I saw clearly how women are the solution to our global problems. I kept hearing the drumbeat that women were making around what was possible for the earth, communities, and future generations. I was stirred by the feeling of the ground-spring of energy that would be available if women could strategically unite around their efforts and wisdom and solutions. I believed we could create a space to meaningfully and authentically hold women's leadership in a conscious, unwavering, and powerful way.*"[3]

WEA's programs are based on the belief that when women thrive, communities, the environment, and future generations thrive. Melinda and Amira are changing the world, even in ways that they may not be aware of. Melinda and Amira are "warrior bodhisattvas" for the Earth. Many visionary warrior bodhisattvas like them are benefitting our world, women and men alike. They work with loving intention.

What might be achieved if women were fully empowered? Not just politically and economically, but spiritually empowered? In 2004, thirteen indigenous grandmothers from around the world arrived in Phoenicia, New York.[4] The grandmothers represented indigenous peoples from the Amazon rainforest; the Iroquois Confederacy; the Arctic Circle; the plains, forests, and highlands of North America; the mountains of Tibet and Oaxaca; and the rainforest of Central Africa. They came from the four corners of the planet to spread healing, education, and prayer for the Earth and all of her inhabitants. The reason they went to Phoenicia was to meet in-person from their far-flung homes and share their deep concern about the destruction of the planet, wars, poverty, and the threat of nuclear weapons. Their goal was to unite as one, speak with one voice against the rampant culture of materialism, and the need to protect mother Earth.

Women elders traditionally were the guardians of the physical, emotional, and spiritual health and survival of the family and the community. This is why many indigenous peoples seek advice and guidance from the elders in the community. It is often the grandmother's role to care for the seventh generation. In ancient Native American lore and beliefs, one must take into account seven generations into the future. Thus, one would not pollute a river or destroy a species that could be a loss to one's descendants. For example, in the classic Yupik story of the greedy and gluttonous Amik, his grandmother comes to his rescue.

As we saw earlier, in "Through the Eye of the Needle," Amik's grandmother has been waiting for Amik to return.[5] She senses that all has not gone well. As she weaves, she feels the Earth begin to tremble under Amik's approaching footsteps. She goes outside and is shocked to see, towering above the trees' tops, Amik's enormous head. "*This is what Amik has now become,*" she says sadly to herself. As the monstrous Amik comes close, towering over his grandmother, she shouts, "*You cannot come home like that, Amik.*" Amik is stunned as he hears these words, and remorse fills his heart as he lowers his huge head in sadness. "*Amik, do you realize the great suffering you have brought through your greed, the harm to our brothers and sisters, the animals and the fish, and the birds?*" asks Grandmother. A huge tear as large as a bucket of water splashed upon the ground.

When Grandmother realizes that Amik is truly remorseful, she decides to help him. She walks back into her roundhouse and puts on her ceremonial shaman robe made of animal skins and feathers. The robe is long and hangs below her knees, going almost to the ground. She readies herself with intention, picks up her special needle, and comes back outside. Holding her sacred needle high above her head and speaking loudly for Amik to hear, she says, "*Amik, you must go through the eye of the needle.*"

Amik is stunned. He wonders how anyone as large as himself could fit through the tiny eye of Grandmother's needle. Yet Amik knows his grandmother is powerful, and he trusts her and her wise ways. As Amik surrenders to his grandmother's wishes and gathers intent to enter the eye of the needle, a loud humming sound escapes from the needle. Soon, a spiral of energy begins to take form from inside the eye of the needle.

The spiral grows in size and soon completely encompasses the monstrous Amik. Then the sound gets louder and roaring winds begin to surround the spinning spiral and enter into the eye of the needle. Amik's body is miraculously sucked into the spiraling winds and pulled into the eye of the needle, as each molecule, first swirls in the inner cyclone of the spiral and then is reduced to a tiny size.

Once fully inside the eye of the needle, Amik's consciousness expands greatly. He sees the entire cosmos and billions of beautiful spiral galaxies of every color and hue. Amik is awed at the cosmos' beauty and feels the Creator's compassionate heart, which grows his wonder. Then, abruptly, his consciousness is pulled back to Earth as Amik "becomes" every creature that he has wrongly harmed with greed in his heart. Once within the spiraling eye of the needle, Amik experiences time at a vastly different rate, with years condensing into seconds and the past, future, and present combining. He lives the life of those he has wrongly killed. He experiences everything they have experienced from their birth to their death. He feels the fear and suffering of their pain and death at his ungrateful hands.

In one instant, Amik is the tiny naked bird baby within its egg, waiting to hatch with its alarming bird mother chirping as Amik's hands grab it from the safety of its nest. In another instant, he is a salmon gasping for breath as it suffocates upon the sand, knowing it cannot mate and lay eggs to start another generation. Amik is then the whale that was born free in the great vastness of the sea, now pregnant and about to give birth to its first offspring, but both are needlessly killed by Amik for the sake of his boundless appetite. Over and over again Amik experiences each life and each death. This experience truly humbles Amik as he realizes the inherent dignity and longing for life in all of those creatures. He realizes that they are much like himself.

As soon as the last life is lived, the spiraling vortex begins to take Amik up and out of the eye of the needle. As Amik exits the needle's eye, all the creatures he has eaten are brought up with him. The many animals, fish, and even the river flow from Amik's mouth, fully whole. Each soon finds itself back where it was before Amik killed it. The mother whale, and unborn daughter are back in their vast ocean home; the salmon are back in their streams; even the delicate bird eggs appear back in their nest. Amik even hears the mother bird's cries of joy at finding her eggs safe and sound. Now, her joy is his.

Amik then fully emerges from the eye of the needle. He then stands humbly in front of Grandmother in his normal young boy's body. He thanks his grandmother and tells her how sorry he is and how he now has another chance to be a "real person: and honor the ways of his people." Grandmother is silent and then says, "*The animals, the fish, the river, and even the tiny birds have filled you with their ways of knowing. You now know of the sacredness of life, and because of this, you have become wise among men.*"[6]

The story shows us that as humans, we have the choice to live for others, not just ourselves. Then the chaos we have caused can be transformed. This requires honoring the feminine qualities within each of us, qualities such as compassion, intuition, gentleness, cooperation, and leading from the heart not just the head.

Just like Amik's grandmother, the thirteen grandmothers have been called to travel thousands of miles from their homes to bring us all through the "eye of the needle." They believe that humanity's spirit has been corrupted and that we have lost the most important awareness of all: that all life is sacred. The grandmothers are actively participating in redeeming humankind. They refer to this time as the "Purification Times." They see that many changes will be taking place to cleanse the materialism and greed of our age. They seek to deepen our appreciation of the divinity in all things so that we, too, may be transformed.

Other indigenous women are stepping forward as way-showers. Native American women from the Standing Rock Lakota-Sioux Nation took action to stop constructing the Dakota Access oil pipeline. The women elders gathered their people, men and women alike, to rise and defend their sacred land.[78] Standing Rock Lakota-Sioux Nation women abstain from using physical violence and ask all to avoid thoughts of violence. They believe that even contemplating aggressive actions against their oppressors could make matters worse. Hundreds joined this effort peacefully, even though they were cursed, beaten, and attacked by dogs.[9] Many were injured. The water protectors (as they called themselves) were willing to risk life and limb to stop their sacred sites' desecration and protect their primary water source. (One woman may have lost her hand after militarized police threw a grenade at her; another lost her sight when a pepper spray can was thrown in her face.)[10] Yet, the women and men peacefully continued their resistance.

Sharon Kring, a journalist and filmmaker who stayed at Standing Rock and produced a film documenting their efforts described the Standing Rock women as an example of uniquely feminine leadership.[11] Kring admired their fortitude and their reasons for taking action. "Unci Maka is Mother Earth," explains a Standing Rock mother of four in the film, who got involved in the action. *"The land is not a resource. It is an entity with whom you have a relationship, and you respect, like your grandmother, mother, and aunts. We have to be good stewards. We have to take care of the land. We have to feed her and offer her prayers. In return, we are blessed with good health. Everything comes full circle."*

Kring chronicled the Standing Rock response as largely a feminine approach to protection. She defined this method as having three aspects: (1) The women's involvement was motivated by a desire to serve the greater good. (2) The women exhibited a diplomatic balance of fierceness and gentleness. (3) The women had a sense of responsibility that sustained their strength and fortitude when confronting wrongdoing.

This uniquely feminine leadership attitude is the way of a "warrior Bodhisattva." The "warrior Bodhisattva" stands strong by modeling ethical actions, such as protecting people and Nature, abstaining from physical violence, and calling upon others to join them. While warrior-like characteristics can be found among environmental activists, what is rarer is to be able to be loving and compassionate towards those involved in harmful industries. Too many "younger brothers" are caught up in wrong-headed socio-economic trends, or a failed religious fervor that results in harm to others. Yet, they may someday change. Warrior bodhisattvas remain committed to stop harmful actions and industries and to find ways to engender wellbeing. Many remain compassionate even in the face of terrible odds. They understand that people can change, and thus they focus on actions without damning the individual. They understand that someday, that person destroying the environment may change and become an ally in protecting the planet.

Sometimes, women's leadership is highlighted by a crisis. In the case of the highly contagious COVID-19 virus that literally shut down the world, it was disproportionally the women leaders and heads of state that most readily made life-saving decisions. Countries with female leaders like Germany, New Zealand, Iceland, and Finland made the bold and unpopular call to shut down and shelter in place days or weeks before most male leaders.[12] The mayor of San Francisco, London Breed, who is the first black woman ever to hold that office, also reacted quickly to shut down, even though she was criticized for doing so.[13]

These leaders are less interested in promoting themselves and more willing to make controversial choices to protect the public from harm. Study after study has confirmed what the recent COVID-19 emergency has proven. Female leaders manage risk differently than male leaders. A famous study in 1994 found that women and people of color perceive the risks of health and technology hazards much higher than do white men, who are much more likely to take a risk that may result in ill health or even death.[14]

Just last week, Jacinda Ardern, won reelection as the prime minister of New Zealand. Under her leadership, the Covid 19 virus was all but eradicated in New Zealand. Today, while much of the world is still experiencing lock downs, shutdowns and mask wearing, New Zealand is in a universe all its own, with no need for these types of restrictions. Even before the pandemic, prime minister Adern, clamped down on racial and religious discrimination, in her reaction to the Christchurch mosque attack by an Australian national and white supremacist, that killed 51 worshippers.

Arden wore a headscarf when comforting Muslim victims and mourners and cried with them. She offered to pay for all funeral costs, and forcefully pushed new gun control measures forward through parliament banning military style assault weapons. Her approach is a distinctly feminist approach to terrorism. A distinction can be drawn between her approach and the approach of other world leaders, most of them male who have "met" terrorism with mass surveillance, repression of civil liberties, strengthening law enforcement, and calling for a war on "terror." Given the relative social calm, and lack of retribution, in the wake of this massacre, Arden's leadership plainly worked!

There is a growing movement happening in many parts of the world to promote women and their values. A truly historic moment took place on January 21, 2017, when five million women from seven continents participated in history's largest public demonstration. There were 673 Women's Marches worldwide, including 29 in Canada, 20 in Mexico, and Antarctica.[15] The Women's March in Washington, DC, alone drew 500,000 people. Not only was this the largest international gathering of people for a common cause, but it took place without a single act of violence. As women move into positions of power and influence, we can hope that our species' willingness to risk it all will be tempered. As women step out powerfully in self-expression and leadership, the qualities honored in goddesses of old are being brought to the fore in the life of common women and into the world.

17. Spirit Horse

When I bestride him, I soar; I am a hawk; he trots the air. The earth sings when he touches it. The basest horn of his hoof is more musical than the pipe of Hermes.[1]

William Shakespeare

The single-engine plane glided at three thousand feet above the Texas coast. The blue ocean below glittered with flashing diamonds of sunlight. What I love most about flying is being able to see the Earth from on high. Envisioning the land and the water from that great height lifts one into a new reality. From the sky, the Earth's body is a vibrant being, resplendent in her garb of green plants, blue oceans, and reddened deserts. This view, previously only available to the great raptors, is now a common one for humankind. Seeing the Earth from the sky reveals an illuminated landscape. Yet, it was a landscape even our ancestors had access to.

The ancients "knew" our Earth long before planes or NASA. They gained their knowledge not from launching fuel-burning rockets but by listening to the Earth. They connected to her consciousness, her very soul. They lived in a completely different world, a world that is a part of an animated, wise, and compassionate cosmos. Their knowledge influences their day-to-day reality and allows them abilities and visions; modern man can only dream of.

By downplaying a purely personal sense of place and space, they acknowledged their relationship to something greater than themselves. Their "relational paradigm" fueled their beliefs and informed their societal and individual actions. For example, ancient Picts/Celts were in tune with not just the Earth but the cosmos. They built their roads and structures in alignment with the heavens above.

Unlike the Romans, who feared the forests and focused on expediency, the Picts traveled a Divine Map.[2] This map encompassed three realms, considered separate but interconnected kingdoms: Middle Earth, where humans live; the Lower Kingdom was another, while the Upper Kingdom, the Heavens, was the realm of the Greater Gods. This included the planets and the sun, whom the Picts/Celts recognized as sentient beings along with the Earth.

They created roads and pathways in the tracks of the sun's summer and winter solstice in Britain, as did the Celtic Gauls on the continent.[3] Their buildings and routes in the Middle Kingdom where they walked are based on the stars and planets above. Their in-depth knowledge of astronomy informed the placement of their standing stones and places of worship. This had the effect of sanctifying their day-to-day lives with an awareness that they lived within a larger, inspired cosmos.

Julius Caesar said about the Gauls/Celts that: "*The cardinal doctrine they seek to teach is that souls do not die, but after death pass from one to another; and this belief, as the fear of death, is cast aside, they hold to be the greatest incentive to valour. Besides this, they have many discussions as touching the stars and their movement, the size of the universe and of the earth, the order of nature, the strength and the powers of the immortal gods, and hand down their lore to the young.*"[4]

Without advanced technology, the ancient Celts watched the stars for generations and passed down their knowledge. Much of what the ancient Celts knew thousands of years ago is more accurate and insightful than what we can teach children in school today. In the case of the "unseeable", spiritually evolved leaders like Druids and shamans could go within and connect to a larger "cosmic mind."

On that day, flying the Katana to visit the Audubon sanctuaries, I was a relative innocent in understanding how drastically the world would change. Few at the turn of the last century could have imagined the impact the human race was to have on the planet. Today, with the specter of climate change looming large, many conservationists see their actions as band-aid remedies and possibly ineffectual in the long run.

As the ocean glittered beneath me like a million diamonds, I recalled a dream. At the beginning of the dream, I am standing with my older sister, Jo, on a vast, open plain filled with grasses and wildflowers as far as the eye can see. With the light of the setting sun shining on her face, my sister points to something in the far distance. "*It is a horse, a runaway horse*," she says quietly, but with great emotion.

Looking out into the distance, I see only the empty expanse of a wide-open plain, flowers slightly moving in the breeze, and tall mountains in the far distance. I am perplexed and turn to reply that "*There is no horse*," when in a dizzying instant, the furthermost point on the horizon jumps forward, making the very far seem now all too impossibly near. Although initially disorienting, the minute details of that once-distant place and the abrupt shift in visual perspective is wondrous. At the furthest reaches of the plain, the tips of the golden grasses filled with tiny seeds can be seen, as well as individual petals of colorful wildflowers.

Everything seems close, impossibly close. I realize that I cannot see the horse in the far distance because I am looking directly into its eye. Then, a woman's tiny image appears in the black pupil and rapidly grows as if my attention has called her to the forefront. As she comes into full view, I am shocked to discover that the woman in the eye of the horse is me. My consciousness then shifts sharply, and I am no longer a separate observer. Instead, I am inside the horse's eye and, stranger yet, I am the horse's rider at the same time.

With my hands wrapped in the horse's silky white mane, the dream horse moves weightlessly over the land as I have the odd experience of two landscapes simultaneously: the outer plain where my sister and I stand together and the other, more mysterious, landscape from within the horse's eye. Here distance and time seem non-sensical, and the colors appear vibrantly surreal. Simultaneously, somehow everything seems connected to everything else. Meanwhile, the flowers and the grasses vibrate and emit different tones like an invisible heavenly bell-like chorus. It sounds vaguely familiar as if I have heard this beautiful music before.

As I hold fast onto the horse's mane, I wonder if this could be a lucid dream.[5] Leaning to one side, I hold out my hand to touch the tips of the grasses that go by like waves upon an undulating ocean. The tips bounce off my outstretched fingers in a staccato motion as we fly with dizzying speed. In the internal landscape, the sounds of the struck grasses peal in excitement.

The dream horse is running towards something yearned for, something beautiful. It seems almost in our reach when the dream horse suddenly leaps in one huge stride away from the Earth. I close my eyes and hold my breath, waiting for the touchdown. Yet, the familiar thud of hooves hitting the ground never comes; instead, I have the sensation of freely floating. Weightless the Spirit Horse and I travel upwards, away from the Earth, above the clouds, past the blue sky, past the black and dark blue of the furthest atmosphere, and onwards into the dark deep space of the cosmic mother.

Feeling for the horse's mane in the dark, for a better hold, as soon as my fingers reach for the silky whiteness, it dissolves, like a mist between my fingers, until there is nothing there. I, too, am dissolving into that ethereal mist. Perhaps the horse and mine's disappearing molecules have co-joined together because the horse and I are no longer separate beings. Our combined consciousness, *Homo sapiens*, and Equus have united as one.

Then, there is no horse; there is only light. A luminous, radiant light that fills everything. A light so bright it should have burned our eyes, but instead "embraces" us, or what was left of us. Perhaps, the light is what emanates from our molecules surrendering in a flux of pure energy. Liberated from the physical, we float freely in a silent, peace-filled Universe, innocent observers of all that is.

As I float freely in the vast expanse, two beautiful enormous planetary orbs appear as living conscious beings. One of the planetary orbs engenders a deep sense of familiarity. The other, a radiant sun, shines a golden light, with rays so fine and subtle that they only illuminate the face of the planet, not the starkness of the void. From some hidden recess in my consciousness, the word "Earth" arises. Yet, the planet I call Earth appears unusually hard and metallic. The forests, oceans, deserts, and cities all seem subdued under this veil of hardness. As my heart fills with sadness for the melancholy planet and its inhabitants, a shock of cosmic light suddenly bursts forth from the deepest recesses of the Universe.

A vast, bright ribbon of light, thousands of times brighter than the sun and a whiter hue, shoots forth like an arrow from the bow of some great cosmic being. The ribbon of light gracefully curves this way and that as it encounters the subtle fabric of space as it folds like a piece of ribbon candy. The ribbon slightly touches the Earth's surface, and where it touches, a white glow permeates the metallic surface and starts to spread slowly. As it spreads, the very character and substance of the planet becomes transfigured, transmuted. I wondered what it all meant and why the Spirit Horse had brought me here.

It was as if a window had opened upon a strange reality. A world where the Earth's spiritual and energetic aspects, the cosmos, and every living thing were set plainly before me. The vision was so astounding that I began to wonder whether I had dreamed of the Spirit Horse or if it had dreamed of me. I witnessed this world-changing event from an objective perspective as if I was witnessing an event unrelated to myself. I had passed beyond individual human consciousness and comprehension and felt outside of my own body.

As I observed the cosmic ribbon transforming the world, I considered the catastrophic impact to all life upon her surface. As this thought arose, a divine cosmic presence, whose own consciousness dwarfed my own, explained that it was a much needed spiritual "*softening*." The cosmic event was not just a physical change and transformation, but a massive spiritual realignment. It was only later that my mind came to understand the meaning of the vision, and for that to unfold, I needed to embrace the mystical, invisible world.

V. AETHER

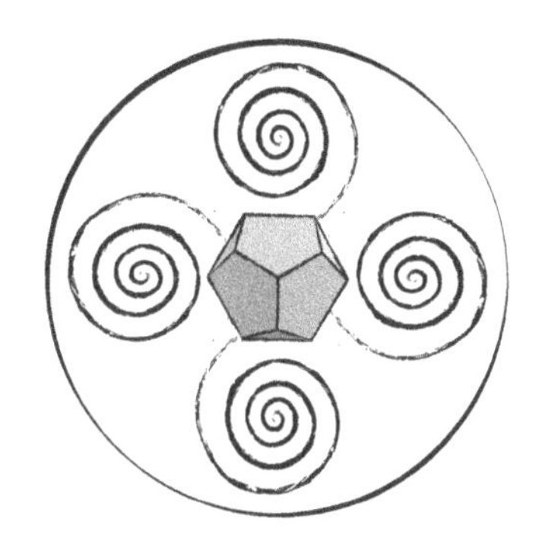

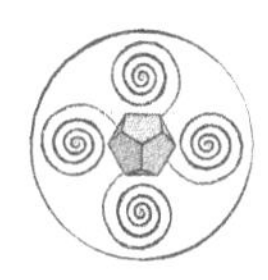

18. Mind Over Matter

Mind over matter.

SENSEI HIRAOKA

The universe begins to look more like a great thought than like a great machine.

JAMES JEANS[1]

"*Mind over matter*" was an often-used expression of my esteemed teacher, Sensei Hideaki Hiraoka, Ph.D. I studied with Sensei over a seven-year time period and learned a great deal from him. Most of what I had learned about subtle energies and the invisible realms had come from the trees, the animals, and the land. Yet, here was an exemplary teacher from a wholly different culture who helped me understand cosmic energies and apply them for humankind's benefit.

Sensei Hiraoka followed in his family's King of Samurai tradition from his mother, Tamano Hiraoka. Samurai were the nobility class of medieval and early modern Japan. While they are frequently depicted in TV and movies as masterful warriors, the tradition encompassed a far greater role in their community. The life of a samurai was not only about physical prowess but also about spiritual development and moral responsibility.[2]

Young samurai were raised with an emphasis on self-improvement and care for tradition and country. Samurai thus held an important position by seeing to the welfare of members of the community. The samurai political structure slowly died out in the 1800s and then ended abruptly at the end of the Second World War when all samurai had to give up their swords. While the samurai no longer have a formal role in Japanese society, a recent study found that samurai descendants still hold influential Japanese society positions.[3]

Tamano was a famous energy healer and wrote several books on the topic on the mind's healing powers. During World War II, and after nuclear bombs were dropped on Nagasaki and Hiroshima instantly killing over 230,000 civilians and causing birth defects on thousands more, Tamano did her best to attend to hundreds of injured and ailing people. She would take her son, Hideaki, only nine years old, with her as she went door to door as many hospitals were destroyed and many nurses and doctors were killed. No doubt Hideaki learned a great deal about healing but chose a more scientific and technological career path in his early adulthood.

A true renaissance man, Sensei held a doctorate in engineering and a doctorate in science. He taught in the engineering department at UC Berkeley and won a Fulbright scholarship in his twenties. He was hired as a senior engineer to design nuclear power plants and managed a team of over 270 engineers. Sensei held multiple black belts, obtaining his first Judo black belt at 13 from the Police Academy School in Japan. In his later years, he was a judo and karate teacher and taught thousands of adults and young people over 30 years. This was when Sensei was featured in the 1980s TV program *That's Incredible!* breaking 2,600 pounds of ice with one kick. After obtaining superhuman physical power and swiftness, he turned his attention to working with Ki-energy (which he referred to as Cosmo Power) and devoted himself to healing difficult cases, such as curing people of addictions, chronic illness and teaching people how to increase their own Ki-energy.

Over the years, I observed him heal many people as well as to conduct superhuman feats. For example, Sensei could heal people instantly, even those who had been bedridden or in wheelchairs for many years. He would apply his high-energy touch, and they would be walking within minutes. He could even heal people on the other side of the planet with a simple phone call. There were always people coming to Sensei for healing. One or two evenings a week, myself and two other students would be allowed to perform healing under his watchful eye using the Cosmo Power techniques to shift and change each other's energy.

We would begin around 6:00 PM and often would not finish until well past midnight. Sometimes we sat with large medical texts laid before us as Sensei discussed the human body's finer aspects. Other times Sensei would have us follow an exercise routine that he created to emphasize maintaining the body physically with stringent exercise. Other times he would discuss healthful foods and supplements.

Most of the time, though, we "worked" and "played" with energy. We healed people. We practiced viewing auras and energy fields. We learned to "project" our energy towards another person. We discovered methods such as applying positive energy that would uplift another person from 15 feet away or how to apply energy that could "lock" up their muscles, making it difficult for them to move.

Sensei enjoyed displaying the powers that flowed into him from what he termed "the Cosmos." He had and could display what some might consider to be supernatural powers, and fortunately chose to use these for good. Sensei extolled the virtues of beneficial cosmic power or energy and enjoyed applying it so that people could have better health and greater cognition, run twice as fast, or jump twice as high. Sensei understood that sometimes people needed to see and experience how cosmic energy works to believe in it.

He had a great sense of humor and sometimes enjoyed pranking people and surprising them with his abilities. For example, Sensei could apply energy to one's brain to not be able to recall or say their own name. Or he could make it difficult for people to move their feet by "freezing" up their muscles. He could also do the exact opposite and supercharge one's brain to be more efficient at problem-solving or charge one's leg muscles so they could run faster. He understood that energy was essentially neutral but could be shaped to be beneficial or harmful.

Sensei taught that the body could heal itself as long as the frequency of the cells are in an optimum state for health. He designed and developed a special machine that could read the subtle frequency of cells on any part of the body and quickly identify an unhealthy frequency. While Sensei could distinguish this on his own, he wanted others to be able to tell when cells fell out of a healthy frequency range with the help of a machine.

Sensei believed that our state of mind was directly affected by frequencies and could be altered through thought alone. He once helped a widow mourning her husband's death for many years by "lightening" the hold of anguished thoughts from her brain. Sensei referred to this as "*cleaning up the brain,*" and he encouraged us to do this daily to arrest the degrading influences of negative thinking and modern-day stresses. He saw these "*degrading states of mind*" as a hindrance to emotional, psychological, and psychic development. A troubled mind cannot tap into the subtle cosmic energy that surrounds us.

Astonishingly, Sensei's body maintained a protective invisible shield that could, when danger loomed, make it impossible to penetrate within three to eight feet of his body. He exhibited this "shield of energy" when he taught judo and karate. There is even a YouTube video of him repelling Willy Fabros, the Filipino karate champion, and causing him to be thrown backward by Sensei's invisible but powerful shield.[4] According to Sensei, this shield arose automatically as a protective layer when under attack, without him needing to think about it or do anything.

One day, Sensei shared with us the story of how, when he was teaching at Berkeley at 27 years old, his car broke down on the outskirts of town. As he was pushing it out of the way of traffic, three large men approached him, saying they wanted to help. When they pushed his car into a deserted lane, they showed that they intended to rob him. They probably thought that Sensei would be an easy target since they towered above his five-foot, two-inch frame.

One of the robbers approached Sensei from behind, ready to land a blow on the back of Sensei's head. Much to the robber's astonishment, his hand collided not with Sensei's head but with his energetic shield, sending shock waves through his body, freezing his muscles, and causing him to fall helpless onto the ground. As he lay there paralyzed and unable to move, the other two robbers, confused and frightened at their friend's plight, ran away. They learned that Sensei was no ordinary man.

On rare occasions, Sensei would bring out his swords and allow some of his longtime students to "test" his energy field. On one of these occasions, two other students and I would charge Sensei with swords. Running towards him from three directions, when we collided with his invisible force field, it felt like hitting a brick wall. Not only does the force of the field knock one back and repel one at the same level of force brought to bear, but Sensei's energy field would also cause one's muscles to freeze up. One student was so overcome by the power of the shield that he went unconscious for a few moments. While Sensei's abilities were fascinating and educational, he knew that the most important work that one can do with cosmic energy is to heal others. He devoted the majority of his time to this work.

One of Sensei's teachings' foundational tenets is that gaining mastery of one's own brain wave frequency is key, with the alpha and theta states being the most important. Sensei believed it to be essential to be in the state of *satori,* which he described as being in the presence of the "state of nothing." Satori brings one into sync with the cosmos. Sensei said that once in this state, one's energy, intentions, and thoughts travel faster than light and influence the outer world. Also, one can tap into the constant flow of cosmic energy and be continually "recharged."

~

The human brain generates electromagnetic frequencies that operate different modes of consciousness. A brain in a state of relaxed wakefulness and awareness ranges between 7 to 14 hertz, with 7.8 to 8 hertz being an ideal range, while a brain actively computing or processing technical information operates at a higher frequency of 14 to 40 hertz, known as beta. When alpha waves cascade throughout a person's cortex, he or she will become calm and centered yet highly alert and aware. Alpha state is a state of peak mind/body/spirit integration for the simple reason that the mind achieves overriding coherence between the inner world and the outer world. Research overwhelmingly indicates that in the alpha state, human beings not only experience deep relaxation and a feeling of well-being, but their healing and psychic abilities are enhanced.[5]

This state of being can be induced through meditation and the conscious control of brainwaves. Sensei explained that when we trained our brains to pick up more subtle energies, more easily accessible in the alpha state, we were accessing a higher level of consciousness. He also explained that this heightened consciousness had little to do with "intellectual" knowledge. While Sensei largely confined his teachings to focus on the more practical applications of "cosmic power or energy" to affect the material world, these sacred aspects greatly appealed to me.

Over a decade before meeting Sensei, I had discovered the meditation method introduced to the West by Paramahansa Yogananda. Paramahansa Yogananda was an Indian guru who introduced millions of people to the teachings and methods of Kriya Yoga. He came to the United States in the 1920s and stayed for over 30 years to bring greater spirituality to Western society, emphasizing material growth.[6] He is considered to be the "Father of Yoga in the West." His book, *Autobiography of a Yogi*, sold over four million copies, with Harper Collins listing it as one of the "100 best spiritual books of the 20th century."[7] Before his death, former Apple CEO Steve Jobs ordered 500 copies of the book for each guest at his memorial.[8]

Initially, every morning and evening, I would sit on a small bench outside our apartment and meditate. While at first, it seemed hard to sit quietly and still, I soon fell into a pattern of being able to enter into a meditative state quickly. When this happened, all sorts of beneficial things began to occur, including sleeping peacefully and soundly. Before I routinely meditated, I could not sleep several nights a week as my mind went over and over trying to solve daily problems. Through meditation, I could also access a door to possibilities not available in an "ordinary" state of mind. Meditation provides entry to a vast cosmic consciousness of wisdom and love. Over the years, I began doing most of my meditations out in nature. I developed ways to accentuate the meditative state out of doors by identifying places in the land with beneficial energies. Meditation brought a new and higher level of perception that enriched my life. It is a practice available to everyone to "train your brain and nervous system" to access enhanced consciousness.

While my brain might have been in an alpha state when I was meditating in Nature or in a trance-like state, Sensei's brain was always in a state of "alpha" or satori (nothing). Japanese scientists and medical professionals conducted tests on Sensei. They discovered that not only was his brain in alpha state at all times but that the portion of his brain in alpha was ten times larger than the average person's. This is why he was capable of performing instantaneous healing and other unusual physical feats. By maintaining his brain's health and physical health through exercise into his later years, Sensei was even able to beat Japan's Olympic trial record for the 100-meter dash when he was in his seventies. The judges would not consider entering Sensei due to his advanced age, but it opened their eyes to what a human being is capable of.

World-renowned psychics and healers also operate from the alpha frequency when receiving intuitive information or relaying healing energy. For example, a famous psychic named Peter Hurkos, who was especially gifted in finding missing people, obtained his "special gifts" after falling off a ladder, received a brain injury, and was in a coma for three days.[9] Upon regaining his consciousness and health, he was able to see past, present, and future. According to Hurkos, he would see pictures in his mind like a television screen when he touched an object. It has been shown by way of electro-encephalography (EEG) machines that Hurkos was most successful in locating missing people when his brain produced low-frequency waves at approximately eight hertz. From 1956 to 1958, Hurkos was tested, under tightly controlled conditions, by Andrija Puharich, MD at his Glen Cove, Maine medical research laboratory. The results proved that Hurkos' psychic abilities were real and consistent.[10]

When we access alpha brainwaves, we are in communication with the life force itself. In 1954, physicist Winfried Otto Schumann measured the frequency of the standing waves at the Earth's surface and in the surrounding cavity of space between the planet and the ionosphere. He found that they matched the brain's alpha state of 7.83 hertz.[11] These low-frequency waves not only surround the planet, blanketing all life forms upon its surface, but they penetrate below the surface. It is no accident that the alpha state matches the Earth's predominant frequency. According to Herbert Konig, Schumann's successor at Munich University,

There is a harmonic relationship between the Earth and our mind/body. Earth's low-frequency iso-electric field, the Earth's magnetic field and the electrostatic field, which emerges from our body, are closely interwoven. Our internal rhythms interact with external rhythms, affecting our balance, REM patterns, health, and mental focus. SR waves probably help regulate our bodies' internal clocks, affecting sleep/dream patterns, arousal patterns and hormonal secretion (such as melatonin). It becomes obvious that in deep meditation, when waves of alpha and theta rhythms cascade across the entire brain, a resonance is possible between the human being and the planet . . . Perhaps the planet communicates with us in this primal language of frequencies.[12]

When living beings come into sync with the larger song of the earth, they can download cosmic truths, communicate with others across vast distances, and connect with the trees, animals, and the planet. Instantaneous communications can take place between people and even between people and animals and plants. In alpha, we can gain deep insights into ourselves and the world around us. We tap into a universal channel common to all life. Babies and young children spend a far greater period of time in the alpha state than adults.[13] This probably explains why I could "speak" to the animals and trees more easily as a child. I was in an alpha state and accessing the universal channel of Nature. Alpha is a doorway that can connect one to all of life and the larger cosmos; it is an entrance to the sacred.

While the alpha state is more natural for children, as we grow, society encourages us to ignore the "information" and sensations that we receive and focus instead on the nuts and bolts of living. This means spending much more time in an analytical beta brainwave state to succeed in school and work. This state emphasizes number-crunching and receiving and distributing information through calculation and memorization. Besides, corporations and the military are using frequency technology for communications and even military operations.

Now that we understand that all living things are made of energetic frequencies, we can comprehend the harm resulting from blanketing the world in artificial electromagnetic frequencies. These frequencies, sometimes grouped under the name of EMF technologies for cell phones and computers, have been linked to low sperm count and even cancer.[14] Startling new evidence suggests that male infertility has grown rapidly over these past 20 years throughout Western society. A 2017 study[15] titled "Temporal trends in sperm count: a systematic review and meta-regression analysis" found that sperm concentration had fallen by 52 percent among Western men between 1973 and 2011.[16]

Today, communication companies have received approval to launch 5G. 5G is the "5th-generation mobile network" meant to deliver higher peak data speeds, ensure more reliability, and expand network capacity to more users.[17] Unlike 4G and all previous networks based on radio waves, 5G resonates at a higher frequency beyond radio waves into microwaves. This will be the first time that the entire Earth and all living things will be subjected to this type of microwave technology, and yet no safety or human health studies were required or performed before launch.

Radiofrequency radiation exposure from the iPhone 7—one of the most popular smartphones ever sold—is one device that presents harmful radiation. For example, one study found that radiofrequency radiation exposure from the iPhone 7 was measured over the legal safety limit and more than double what Apple reported to federal regulators.[18] Also, the FCC's safety guidelines do not measure the full exposure to cell phones since many people keep them in their pockets next to their body, and FCC guidelines fail to require tests that contemplate this reality and instead test phones away from the body. Some people are already developing sensitivities to these frequencies and cannot be near devices that radiate them or get ill.

Harmful EMFs may also interfere with wildlife and insects. According to Lisa Henkes, the author of "Radio Frequency Radiation (EMF) threatens Plant and Animal Species with Extinction,"

"The accumulated clinical evidence of sick and injured human beings, experimental evidence of damage to DNA, cells, and organ systems in a wide variety of plants and animals, and epidemiological evidence that the major diseases of modern civilization—cancer, heart disease and diabetes—are in large part caused by electromagnetic pollution, forms a literature base of well over 10,000 peer-reviewed studies. If the telecommunications industry's plans for 5G come to fruition, no person, no animal, no bird, no insect and no plant on Earth will be able to avoid exposure, 24 hours a day, 365 days a year, to levels of RF radiation . . . These 5G plans threaten to provoke serious,

irreversible effects on humans and permanent damage to all of the Earth's ecosystems."[19]

While there have been fewer scientific studies on the effect of EMFs on human consciousness, it's likely to affect us because we are beings that generate frequencies. If we are subjected to artificial frequencies that drown out the Earth's Natural frequencies, undoubtedly, there will be ramifications.

On the other hand, beneficial frequencies can be supportive and expansive of life systems. For example, beneficial brainwaves like alpha can be compared to a radio frequency; once you have tuned in, you can hear what is being "played" by other living things. It's as if you tap into a whole new world by tuning into the "channel of life." You can then communicate thoughts and intentions along the same channel with ease and influence the world around you. While Sensei was a master at harnessing these invisible frequencies, all people have the ability to use these to benefit themselves and others.

Just as a slight disturbance on the surface of a pond creates ripples that spread across the entirety, your thoughts and intentions can spread like waves throughout the Universe. In this way, humankind can and does affect the world around us. This is exactly what the Kogi have told us as well. *We* can learn to "shape and mold" certain energies and frequencies in our control. For example, as I have shown participants in my quests, it is possible to reduce, and in some cases eliminate, the harmful energy emanating from unkind neighbors, antagonistic people, and even to "calm" the negative frequencies of personal cell phones and computers via intent.

This leads one to ask: What might happen if the human race were to fully tap into our energetic and spiritual natures and work directly with, not against, the forces of Nature? No doubt, we would accomplish feats that appear impossible to many today. According to the Kogi's, using our consciousness in this way is exactly what we are here to do on the Earth.

While Sensei was truly an extraordinary human being, a superman, he believed that all human beings have the potential to cultivate and master their innate brainpower and tap into the infinite cosmic energy. Sensei believed that one must not underestimate one's abilities, and he proved that consciousness could affect the material world daily. He also said that the more one uses one's abilities to benefit others' care and healing, one's mastery over subtle energies would increase. I have found this to be true in my own life and spiritual work.

The most important "take away" is that when we connect to the invisible energies around us, we can tap into our own sort of cosmic "lift." Just as a heavy metal plane can accomplish flying thousands of feet above the earth's surface with help from the invisible element of air, all people have the ability to accomplish things never thought possible by joining with the subtle, unseen cosmic mind. When we tap into the invisible forces inside of us and around us, in essence, we are engaging with a spiritual and energetic "lift." We can then see and experience ourselves and the world from an elevated level of consciousness. We can then heal ourselves and others, solve seemingly unsolvable problems, or simply connect to experience the bliss and peace of living in attunement with cosmic forces. This spiritual "lift" transforms us and makes us greater than we could ever be on our own.

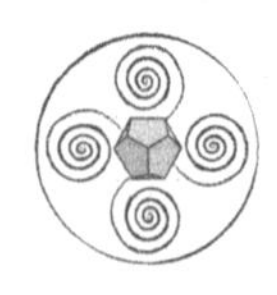

19. Nature Quest

If you sit still long enough
On the forest floor
The universe will approach you
Like a shy animal.

Breathe softly and don't move:
If encouraged it will nuzzle your open hand.

Open more!
Open your heart, your head, your soul.
All doors all bars
That catch and trap and bind
The wild and dreaming beast
That sleeps in you.[1]

MARA FREEMAN

The ancient Celts referred to trees as the "Ancestors of Humankind" While the idea of trees being our ancestors may seem far-fetched, complex life forms would not exist without the plants. Before the arrival of organisms that can conduct photosynthesis, oxygen in the atmosphere was a rarity. When single-cell progenitors of modern-day plants mastered photosynthesis, they increased oxygen in the atmosphere from zero 2.4 billion years ago to 35 percent in the Carboniferous Period (300 million years ago).[2] The newly oxygen-rich environment caused the extinction of many anaerobic and primitive life forms but spurred on the evolution of mammals and human beings.

There are also physical similarities between us and trees. Both have complex "circulatory systems" Besides, Hemoglobin and Chlorophyll have very similar structures. They are remarkably composed of the same four elements: carbon, hydrogen, oxygen, and nitrogen. The main difference is that hemoglobin is built around iron (Fe), and chlorophyll is built around magnesium (Mg). Hemoglobin's primary function is to transport oxygen from the lungs to other parts of the body. It promotes circulation, cleanses the body, increases the number of red blood cells, and therefore increases oxygen throughout the body; in other words, Chlorophyll helps build hemoglobin.

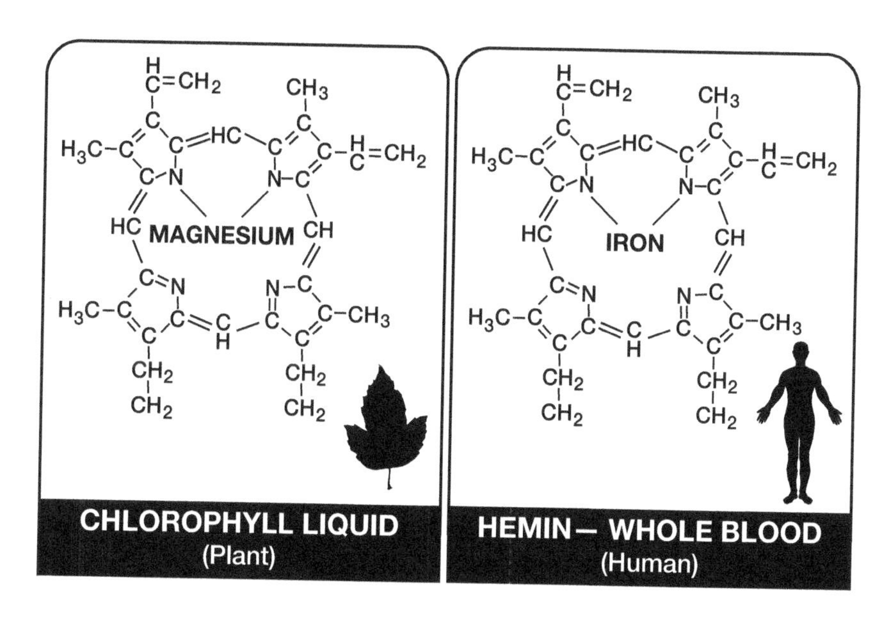

Sometimes physical realities do indeed reflect underlying spiritual truths.[3] The Druids believed that the oak tree was the King of Trees and the axis *Mundi*, the world's center.[4] They saw trees as passageways to a sacred world. The belief that trees are spiritual relatives may seem irrelevant to our lives, yet it still influences us. The commonly used word" "door" comes from the Celtic word "daur" which means an oak tree.[5] It is no accident that a door is something that we can open to step into a different space and place. Similarly, many people bring trees and evergreen branches into their homes for Christmas. They then cover them in lights and beautiful ornaments. These are acts befitting a sacred object. We may not consciously understand the ritual or its ancient roots; it is lock-safe and hidden in our subconscious.

The Celts/Picts, along with other ancient peoples like the aborigines of Australia, were careful not to cut down certain trees. They also mourned the loss of great ones, like those mentioned in the Rennes Dindshenchas, which describes how a "*shining tree-like gold stood on the hill*" whose branches "*would reach to the clouds . . . In its leaves was every melody*"[6] Or the three great yews that took an entire army to cut down. Or the great mother" "Oak of Mugna" said to be thirty cubits wide at its trunk (45 feet) and three hundred cubits tall.[7] The Aboriginal Australians say that native trees are the "kin" of humans, and that some humans can become trees. According to the Adnyamathanha tale from Southern Australia, a man and woman are turned into the wild orange tree (Capparis mitchellii). The Gunwinggu tribe has a similar story of a metamorphosis, where a couple, upon leaving after a quarrel with their families, are turned into pandanus trees (Pandanus spp). The Wuyaliya clan in Yanyuwa Country consider themselves to have descended from the grey mangrove (Avicennia marina).[8] Aboriginal tribes align their clan to a specific plant, then consider it the clan's spiritual keeper. No clan member would ever cut down or destroy that plant for fear of harming their clan. These shared myths and stories among many different cultures show that humankind has a deep spiritual tie to trees.

The "prophetic vision" that I received at Chartres Cathedral hinted at a spiritual calling. Yet, it was ultimately a tree that compelled me to change my life's trajectory and share what I knew of Nature's sacredness with the world.

Many years ago, I had a profound mystical experience as I was coming home late one night after a long workday. I was walking with a full bag of groceries in my arms and a heavy briefcase hanging uncomfortably from my shoulder. As I entered the empty lot next to our building, a soft, golden-white light illuminated the lot. Usually it was so dark I had to be careful of my footing to not trip on the uneven ground. Yet tonight, it was lit with this strange ethereal light. I looked around for a light source expecting to see a newly installed lamp post but could not find any. I then was struck with the fantastic thought that the light might actually be coming from the large tree standing in the middle.

My analytical mind rebelled against this possibility, so I hurried past the tree into our house and closed the door quickly behind me. I leaned with my back against the door with my heart beating rapidly. I recalled a line from a poem I loved by Rainer Maria Rilke, "For *Beauty's* nothing but *the* beginning of terror *we're* still just able to bear.[9] I immediately thought; «*I can bear this*." I put the groceries on the counter, opened the door a few inches, and peered toward the empty lot. Instead of darkness, there it was—the glowing tree, illuminating its surroundings as if it were a golden moon that had come to rest gently on the Earth.

No longer terrified but caught by great beauty, I opened the door fully and walked towards the tree. Standing just under the tips of its boughs, all cares fell away as a deep peace enveloped me. Standing in its soft golden illumination, the blanket of light embraced me. I was filled with the utmost peace, joy, and beauty. I knew that the light was the fire of the tree's divine spirit—the tree's very soul.

When I awoke the next morning, feeling refreshed, the experience with the glowing tree was far from my conscious mind. After taking a shower and dressing, I helped my son get ready for school; made his lunch of peanut butter and jelly, apple slices, chocolate milk, and two Oreo cookies; and packed it neatly into his lunch box. Hand-in-hand, we walked to his school before I headed back to my car to drive the forty-five minutes to my office. At work, not a single thought of the tree entered my mind.

I returned home in the early evening with the late afternoon sun beginning to cast long shadows. Glad to be home, I parked again in the little town center and walked past the peaceful Indian statue towards our home. As I passed the statue, I noticed that a few yellow leaves were resting lightly on his head—a sign of fall. Yet, when I turned the corner the lot was lifeless. There was no tree and the ground was covered with ugly little splinters of deadwood. These shards were all that was left of something that had been indescribably beautiful just 24 hours ago.

My sadness was immeasurable. I let out a loud sob. Then, the horrible sight caused disparate consciousnesses to collide within my being, I felt dizzy and sat down. The part of myself that operated in the mundane affairs of day-to-day life smashed into the part of myself that knew and experienced the invisible and sacred realms. The two different parts were then forged together by the Creator of all things.

I recognized the miracle that had occurred. The tree, knowing of its impending demise, shone forth its light, as if to say, "*I am more than bark and limbs and leaves and roots. I am eternal beauty and wonder. Behold and celebrate what I am and recognize that this marvel exists within yourself as well.*" This singular event transformed me. While my mystical experiences since childhood were kept hidden from the light of ordinary day, the tree put an end to this separate duality in my life.

I was no longer to keep secret my knowing about the wisdom of the trees, the animals, and the Earth. Being a witness to the tree's soul compelled me to speak about, write about, and share these truths with others. While it took me some years to refine and focus my vision and work, the experience led me to write "Partnering with Nature: The Wild Path to *Reconnecting* to the Earth." I also began to lead vision quests and I founded "Nature Quests"™. Working as a quest leader and spiritual teacher has brought me closer to my roots, my authentic nature.

Nature Quests™ are modern-day spiritual pilgrimages. For centuries, human beings have embarked on rites of passage in Nature to gain awareness of the sacred realms and greater self-knowledge. Spiritual journeys into Nature can provide greater life-direction and harmony. These archetypal passages are powerful acts that can change a person's life profoundly. This is why so many myths and stories worldwide have the hero's quest and journey as a central theme. Through Nature Quest,™ people are guided to connect with the soulful and authentic parts of self. By doing so, they gain clarity about their life's purpose. In the hallowed space of the Quest, profound happenings can take place.

While Nature Quests last from 5 to 7 days at a minimum, even shorter excursions into nature can offer dramatic beneficial shifts. Several years ago, while leading a seminar at a national dowsing convention in Arizona, a group of 30 participants followed me outside to a grove of ponderosa pines. While the retreat center's 500 acres of these majestic trees were heartening, it was just a "postage stamp"-sized habitat compared to the 30 million acres that these trees used to inhabit in the Western United States. Ponderosa pines once dominated the landscape, some with trunks five feet wide.

Over 500 years ago, Native Americans had the inspiring experience of spending part of the year in the shelter of these towering pines. Walking in the vast forestlands of ponderosas would have been like walking in an endless green cathedral held up by giant columns of tree trunks towering hundreds of feet upwards. The ancient ponderosa trees large crowns naturally maintained a good distance between each tree. This is why one could see for miles into the distance through the large clearings between each tree. These forests must have been truly majestic up until the twentieth century.

Unfortunately, the logging industry favored these mighty giants, and by the mid-twentieth century, the majority of old-growth pine trees were cut down, never to return. Today, the new-growth pine trees tend to be smaller than their ancestors, though still beautiful. Ponderosa is also kept or planted in cemeteries because they are deemed a species that symbolizes "eternal life"

We walked in silence as our group approached the tree I had selected. Each tree is a unique individual, and some work better than others with certain people and groups. Trees are one of the few beings on the planet on the earth, beneath the ground and in the sky. Thus they" "reside" in three realms and have great mastery in moving energy between them. Human beings also instill and can apply energy to the Earth and other things. I partner directly with the tree(s) to create a safe space and an opening for people to engage directly with its "energy" or aura.

As we stood in a circle around the towering ponderosa pine, the soft wind was filled with the cleansing scent of pine. The clean scent hinted at its healing ways. Inhaling the pine tree scent comes with a healthy dose of negative ions, and negative ions uplift one's mood. The tree's seeds are also edible and medicinally purifying. They have been used by native peoples to clear lung congestion and prevent skin and bladder infections.

Once by the tree, we stood in a circle around it. I guided them in a meditative breathing technique and in bringing energy to the surface of their hands. Within minutes, a pair of tiny bright-yellow and grey Grace's warblers darted in the branches above us, in lyrical song. It seemed as if even the birds sensed the group's harmonious consciousness. I began an invocation and beseeched the tree to commune with each person. I then saw and felt the energy of the tree envelope everyone in the circle. A subtle radiance of golden light expanded and encircled the group. We then tapped into the tree's energy field, and each person experienced an intimate communion with the tree. Some participants felt their hands get hot, others felt their bodies being moved by the tree's energy, and others received visions and answers to long-sought questions.

The participants then shared their experiences. Most were emotionally moved, and some were even physically healed. It is common that many people, after being touched by the soul of a tree, will be in tears from the experience of connecting with a beautiful, sentient being. This is why it is one of my favorite practices, it is one way for people to experience for themselves these beings at a sacred level. In the end, I always bring our healing intentions and prayers to support and bless the trees, the animals, and the Earth. We thank the tree and the Creator for our experiences and send them energy and prayers for their wellbeing. It is spiritual activism.

As we walked back to the retreat center for lunch, one of the participants was excitedly sharing a story with several others. I recalled that he had had a slight limp when we all first walked over to the tree, and had suffered from a chronic foot ailment for over a decade. Now he was pain-free and walking without a limp. He had been instantly healed during our communion and ceremony with the ponderosa pine. A few years later, I ran into him, and he told me how he now routinely interacts with trees and is still so thankful for the healing he received. The truth is that we are meant to engage and commune with trees, animals, and the Earth in this deep way. It is healing for Nature and for us.

It is especially important now that we send our healing thoughts and prayers to uplift and support the Earth, animals and plants. Our thoughts are "things" that powerfully transform the material world. Our intentions matter. This is what the Kogi and many indigenous people have been telling us all along. This is what I learned from trees as a child. The Kogi, who have attuned themselves to communicate with the Earth and Aluna, have perfected this.

Recent scientific studies now show just how powerful our thoughts and intentions can be. Human intentions have been proven to alter DNA. According to a new Canadian study led by Dr. Linda E. Carlson, patients with breast cancer were able to strengthen their telomeres with their thoughts alone.[10] Telomeres are the parts of DNA that prevent chromosomal deterioration. Weakened or shortened telomeres are often associated with cancer, diabetes, heart disease, and high-stress levels. Thus, telomeres are directly linked to life expectancy. It was believed that there was nothing anyone could do to improve upon a damaged telomere in the past. Yet, in the study, patients' telomeres were strengthened and lengthened after patients focused on envisioning them as strong and healthy.

The Transcendental Meditation movement led by renowned quantum physicist John Hagelin has also proven that ones focused intentions and thoughts can alter the material world. In particular, negative human actions, such as criminal behavior, could be significantly reduced by applying focused intention during meditation. In Hagelin's 1999 experiment, 4,000 experienced meditators came together for one month to bring a powerful intention for peace in the Washington DC area. In the same month, violent crimes were reduced by an astounding twenty-three percent.[11] Today, over 100 similar intention projects have been successful in reducing; terrorism, crime, suicides, and even accidents.

The efficacy of these projects has been expertly dialed-in over the years. For example, there must be a minimum number of meditators in relation to the recipient population being helped. That number is the equivalent of the square root of one percent. In other words, one percent of the people of the population sought to be affected. Known as "super-radiance," it is the point at which the joined consciousness of a group of people can change actions and behavior in a specific chosen population.[12] Given today's population numbers, to produce the desired effect for a country the size of the United Kingdom, about 750 people would have to be engaged; in the USA about 1,800 people would be needed; and for the whole world, 9,300 experienced meditators would be needed to have an impact on the remaining seven billion.

Today, meditation and mindfulness practices have skyrocketed. People participate in weekend spiritual workshops or lifelong training. This is a trend that bodes well for the human race. While I believe in the power of intention, I also believe in the power of action. It was not enough for Sensei to enjoy his mastery of utilizing cosmic energy; he also applied himself to healing thousands and to teaching others to carry forth his beneficial knowledge.

Our consciousness affects our relationships with all life on Earth. When we tune in to "Earth's Song" we become greater than the limited and "domesticated" version of ourselves. We tap into Gaia's wisdom and universal cosmic energy. It is then that we can co-create the world with her. By partnering with the plant kingdom, we are uniting with powerful energetic beings that are also capable Earth healers.

~

No doubt I also experience a sacred connection to the trees. The abiding nature of this lifelong relationship was hit home recently when I opened an old box that my mother had packed and put away in the attic many years ago.

Inside one of the boxes was my first "book" written when I was in kindergarten. The book was made out of large sheets of construction paper, with bold, colorful drawings. The story was about a tree. But not just any tree, it was a tree with amazing abilities. The tree was almost burned in a fire lit by people, yet rather than being helpless to save itself from the growing flames, the tree uproots itself and rolls away to a safer place where it then plants itself.

Perhaps I somehow knew that half a century later, millions of trees would be faced with burning from fires. Or that I would be sharing the ways of trees with thousands of people around the world. Yet, it is we who must help the trees in the real life story of the Earth.

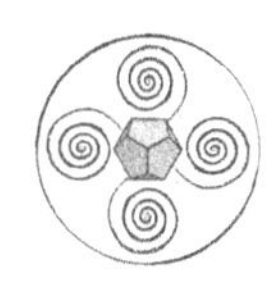

20 Aether

All perceptible matter comes from a primary substance, or tenuity beyond conception, filling all space, the akasha or luminiferous ether, which is acted upon by the life-giving Prana or creative force, calling into existence, in never-ending cycles all things and phenomena.

Nikola Tesla, Man's Greatest Achievement, 1907 (1)(2)

One of my ancestor's most cherished possessions was likely a small, beautifully carved stone sphere. Hundreds of Neolithic carved stone spheres, some over 6,000 years old, have been found in the Orkney archipelago, Northeastern Scotland's mainland, and the Western Isles.[1] Almost all of these round stones are similar in size, with symmetrical patterns and carvings on their surface. Stone monuments and objects found in Orkney offer valuable clues to the way that the ancients perceived the world.

Many Pictish monuments have unusual symbols and animals; the spiral is a common pattern. Based on location and age, we know that the beautifully carved balls are distinct in origin. Many of the spheres are carved with sacred geometric symbols, lines, and shapes. Sacred geometry permeates the architecture of all forms. It is a "blueprint" that shows us the underlying connection between all things and is a continual reminder that everything is part of one great and eternal mystery. Nature reflects and creates patterns and structures from the tiniest particles to the greatest universe. By creating order out of chaos, the Creator infuses the world with intelligence and life; when humankind creates sacred shapes and patterns, we are co-creating with the intelligence that brought us into being.

Many of these Pictish spheres match the shapes of the five Platonic solids that represent the basic building blocks of the Universe: earth, air, water, fire, and aether. These shapes are best known in Western societies from one of Plato's dialogues, Timaeus. In the dialogue, Timaeus explains that an omni-prescient divine artisan creates order out of chaos using just these five elements.[2] The five "solids" are said to represent geometrical archetypes that reveal the perfection of Nature. These perfect geometric shapes are the Tetrahedron, representing fire; the Octahedron, which represents air; the Isohedron, which represents water; and the Cube, which represents the Earth.

In the Dodecahedron's case, it is said to contain all the other four shapes within itself. It is the "womb" of the cosmos, referred to in ancient times as aether. The Dodecahedron is the shape closest to a sphere and considered by Plato to be the most perfect, as it is the most omni-morphic.

While Plato penned the Timaeus dialogue in 440 BCE, the elements were known by my ancestors over a thousand years earlier. The Pictish spheres reflect a high degree of mathematical ability and an understanding of the basic building blocks of the Universe. Many indigenous peoples, including Native Americans, the Greeks, and even the Australian aborigines, understood that the manifest Universe comes from these primary materials.[3]

The fact that everything in the Universe is composed of just five elements may seem impossible and even absurd. Yet, if we look at DNA, the fundamental code of all living things, the natural principle that simplicity can birth great diversity is confirmed. DNA comprises only four chemical bases—adenine, guanine, cytosine, and thymine—yet there are literally quadrillions of different individual life forms! All have different DNA codes using just these four bases to create unique species and individuals.

While we can observe four of these "elements" as physical manifestations, they exist foremost as invisible energetic realities. The Talmud and many other mystical and religious texts speak to this truth by reinforcing that the visible world reflects invisible realities. From the invisible substrate of aether comes frequency (sound) and then finally denser matter that reflects the energetic characteristics of its origin. These energetic or invisible building blocks affect the material world like a puppeteer pulling strings, or a musician playing the Universe into existence.

Ancient peoples associated the elements with different energies that could bring about balance and wholeness. They understood that each element reflected specific patterns of energy that permeated the Universe and each person. Their cosmological world view could be applied to create a sense of harmony within the human being. Egyptian sages believed that by gaining knowledge of the elements, one could understand the very foundation of life itself.[4] In the Mesopotamian mystery schools, initiates underwent sacred rites related to the elements to enhance their innate nature and connect them to a divine source.[5] Hippocrates, known as the "father of medicine," also believed that a patient's state of health depended upon whether or not the elements were balanced within his or her being.[6]

Denise Linn, developed a powerful program known as "Soul Coaching" to align ones inner spiritual life with their external world by tapping into their authentic self. This 28 day program uses the four elements Air, Water, Fire and Earth, to bring clarity, balance and insight into people's lives. By working with the four elements Denise guides people to discover the truth about who they are and what is important to them by connecting them to the wisdom of their soul. I have been certified by Denise as an Advanced Soul Coach Practitioner and I have used her program to assist people in clearing negative emotions and connecting to their spiritual source as a stand-alone program and prior to my Quests. I have found that people who complete this program are much more ready for the deeper pilgrimage in Nature that the Quest offers. The Quests I lead, in a sense, incorporate the final and fifth element of Aether.

As a spiritual teacher, I also work with the five primal, energetic forces of earth, air, fire, water, and aether. For example, I developed the "Ancient Celtic Trilogy Reading," a unique synthesis obtained by merging traditions of knowledge; including how the elements manifest in each person's psyche, native American spirit animal lore, and by interpreting handwriting. It is a powerful tool for self-discovery. The Trilogy Reading reveals the gifts one was born with and how to apply those gifts to the greatest advantage, it also identifies limiting perceptions and habits that may hold someone back from living an authentic life. The readings help people gain a deeper understanding of self, their intrinsic nature, modes of learning, and belief systems. They can help people gain a clear understanding of themselves, what they want most for their lives, and how to best use their unique talents and abilities. Thus, one can work directly with the elements to bring insight, healing, and empowerment.

There are also ways to work with the physical elements that we find in the manifest world. For example, connecting with the earth's energy via the ground or soil can drain off harmful energies and restore harmony to the body. When we have skin contact with the soil or ground, free electrons can be taken up into the body. The free electrons then act as an antioxidant and reduce free radicals that can cause disease and inflammation. Thus, "Earthing," as it is known, can reduce inflammation and pain. Earthing has been proven to "un-clump" and thin the blood to allow nutrients and oxygen to move easily through the body. We also have cords of energy that connect us to the Earth. We can strengthen these cords by sitting or lying on the ground or walking barefoot.

Some years ago, after a spring rain, I walked barefoot far from the hiking trail, following a deer path. Sitting barefoot on the ground in one of my favorite places by a grove of live oak trees. I felt inspired to place my hands into the moist ground and feel it's cool, damp earthiness between my fingers. I felt a coolness move through my body and an old pain in my knee melted away. Each of us has these special places in and on the land that can "fee" our energy and even provide knowledge.

With respect to water, immersing oneself in water is a cleansing and even a sacred act, as exemplified by baptism. Water can draw off negative emotional charges residing in the body. Pure, unpolluted, and living water is the fountain of youth. Consuming a certain quantity of pure water daily for several months will clean the body's organs and skin, clear the taste buds of bad eating habits, and even lead to weight loss. Cold showers and dips into cold lakes, streams, or ocean water can also invigorate our central nervous system and circulatory system to strengthen our bodies and minds.[7]

Water can be like a fountain of youth. For one year, I studied with a Buddhist monk who guided us in a "water diet" Besides keeping a diet of no meat and eating more fruits and vegetables, I drank four liters of water every day before noon. The weight melted off and, my skin and eyes looked clear and bright. Near the end of this process, I saw a beautiful geometric form appear in my third eye area, the pure, luminous color of new spring leaves. It was a sign that I had reached a level of physical purity that also affected my consciousness.

We can affect water with our words and energy. For example, we can write words on the side of a clear glass and this can change the subtle qualities of the water within. I like to write "Love," "Gratitude," and "Abundance" on the side of a glass of water and leave it out over night under the full moon. I will then drink the water the following day while contemplating the word(s). It is one way to infuse your body with good vibrations in alignment with the words that you use. Since our bodies are 70% water, just think how much the words that we use and experience affect us.

With regard to the fire element, bonfires have always been an important part of sacred Celtic rituals and seasonal celebrations. The ancient Celts celebrated a festival known as Beltane.[8] Beltane (or Beltain) was the Gaelic May Day festival. It was usually held on the first of May, or about halfway between the spring equinox and the summer solstice. Special bonfires, believed to increase fertility and provide protection, were kindled. The ancient Celts would walk, dance, or leap around the fires and sometimes even jump over the fire.

The fire would then be shared with the entire community as all the local folk would light a torch or candle from the bonfire to light fires at their homes and their hearths. Colorful yellow flowers were also used to bedeck the doors, windows, and even the domestic animals because they evoked the fire, which was a symbol of the return of the Sun. Beltane was widely observed throughout Ireland, Scotland, and the Isle of Man for centuries and is celebrated on a smaller scale.[9]

Fire is cleansing. Many people have extolled the virtues of saunas and the Native American tradition of the sweat lodge. Routine use of heat in sweat lodges and saunas can help lower infection risk, reduce inflammation in the body, and eliminate toxins. The sweat lodge also has spiritual significance, as it is believed that a purified body is better prepared for the sacred.

Exposure to the rays of our sun affects each person at a cellular level. Our sun, just like our Earth, is a living being, and its energy is working to support and evolve life on the planet. This is just one of the cosmic forces at work to help humanity grow as a species. Sun-gazing is another technique that can be used in spiritual practices. Sun-gazing is the act of looking directly at the Sun at dawn and/or dusk when it is still low enough in the sky that the atmosphere blocks some of its harmful radiation.

Standing barefoot and connected to the Earth, a guided spiritual seeker may stare at the Sun for a few minutes. Sun-gazers claim that they have heightened mental acuity, sleep better, and more energy. However, this practice can be hazardous! If sun-gazing is not done properly, or if the sun-gazer is not appropriately prepared, he or she could damage their eyes and even go blind! This is why all successful sun-gazers have studied with a guru or spiritual teacher for some time and have been properly prepared.

For the element of Air, Air can enlighten and shift our consciousness. Winds carry messages from beyond and can speak to our souls. Pine forests, waterfalls, and beaches where ocean waves crash on the shore are filled with negative ions that can make us feel invigorated and revitalized. Just inhaling the air at these times and places can shift one's mood and consciousness and increase our energy. There are also places on the land where one can experience what I refer to as "land chi." Land chi is where the confluence of wind, land, and water come together in a healthful way to uplift people and animals in its way. It is the equivalent of an "energy river" moving over the land.

Breathing techniques can also help us feel invigorated and alive and/or become more centered and peaceful. During guided meditations, I ask people to do the "six-count breath." The six-count breath is a breathing method where one inhales steadily for a count of six, filling lungs deeply with oxygen, then holding for a moment, and then releasing steadily for a count of six. After even just seven of these breaths (although more is recommended to alkaline the body), the brain begins to shift into a relaxed but aware state of alpha. (People leading stressful lives tend to breathe at a shallow level, starving their cell, body, and oxygen brain. It's important to breathe deeply daily for at least 15 to 20 minutes to ensure that the body gets the oxygen it needs.)[10]

Just as plants gain their energy from the sun, human beings can gain energy from cosmic particles in air and sunlight through special breathing techniques. This is accomplished through the part of the brain known as the medulla oblongata. The medulla oblongata regulates the body's most important life functions, such as breathing and heart rate. The medulla can receive "nourishment" through distinct breathing techniques to recharge the human body. Receiving life energy from the breath is covered in the classic spiritual book "Autobiography of a Yogi" by Paramahansa Yogananda.[11]

The "Autobiography" refers to two women who thrived without food or water for most of their lives. Therese Neumann, one of the women, was a noted Bavarian stigmatist (bearing the crucifixion wounds of Christ). Her food-free life has been studied and documented by scientists and doctors. Giri Bala, a lesser-known Indian Saint, also introduced to the world in Yogananda's book, was tested by the Maharaja of Burdwan and not allowed to eat or drink for two months. She had lived this way since she was twelve years old and asked to have the desire for food removed from her consciousness. Giri Bala was visited by Yogananda when she was sixty-eight years old. He learned that she had never been sick, radiated a healthful glow, and could control her heart and breathing. Both of these women used special breathing techniques to gain nourishment from the Aether.

Breatharianism (replacing traditional food with a type of breathing that provides nutritional support) can help people access the life force, or prana, from the infinite and eternal stream of energy that surrounds us. While it is not recommended that western people cease from eating and drinking, given the demands of our daily lives and society, by relying in part on this purer, sweeter "nutrition", we can release toxicity in the body and fill ourselves with pure, light-giving healing prana. When accessing this "pranic food," it is easy to reduce one's physical food intake and make better food choices.

As a Breatharian teacher and practitioner, I have comfortably reduced my food to one-half of my usual caloric intake for over a year. During this time, I never felt hungry, had lots of energy, and looked years younger. Although I still practice breatharianism for its physical and spiritual benefits for weeks and months at a time, I also partake in traditional eating patterns at other times throughout the year to match societal rhythms.

I teach breatharianism so that others can benefit from this nurturing way of recharging the body and spirit. If humanity supplemented their food intake with breathing practices that provided them with life force energy, all people would be healthier, have more clarity, and our consumption of traditional food could easily be reduced by one third or more. Just imagine a world where human beings do not have to consume as much food! This would solve so many environmental and health-related problems and stop the suffering of millions of animals we raise and kill for consumption.

~

Perhaps the most wondrous element of all is Aether. To the ancient Hindus, aether, or "Akasha" is the fifth physical substance.[12] It is the stuff of which gods and celestial beings are made. It is the stuff of souls. It is immortal, indivisible, infinite, and indestructible. It is the finest and subtlest of all the elements. While the four other elements of earth, water, air, and fire, may be influenced and changed when they come into contact, Akasha remains pure and untainted by the other elements. It is beyond the grasp of the senses yet acts as the medium through which sound travels and creates frequencies. Aether is the medium through which ardent "seekers" of all religious and spiritual paths communicate with "gods" via sound vibrations of mantras and sacred syllables.

The aether is both an element, where all five elements come together, and metaphysical reality. It is where sound, frequency, and vibration live and bring forth and affect the physical world. It is also where spirit resides, and where the unseen influencers of the material world reside. During Nature Quests, I work with both aspects of aether for the benefit of participants. In many cases, people can be healed from physical ailments. One Quest participant described how she healed from chronic pain: "*When we are connected in such a way and let go, then our emotions calm, and our physical bodies are at ease, and that is truly wonderful, as the pain we have been carrying is no more. I've had constant chronic pain in my left shoulder and neck, which completely went away*" Quest Participant 2018

The Quest's main goal is to ignite an inner sense of purpose and allow individuals to reach their creative potential. As western societal limitations relinquish their hold on the human psyche, insightful interactions with the land and the animals take place. According to Ingrid, who attended a Quest, I led to Orkney in 2016; "*Although my feet have barely touched the ground since I arrived back, my heart remains light, and my thoughts are often in the north. I've come away with a much deeper understanding of myself, and a stronger idea of my potential*" *Quest Participant 2011*

Similarly, Heather, who attended a Vision Quest in California, explained that; "*The Quest has helped me to reconnect with my life's purpose. In addition to accomplishing major changes in my personal life, I am now working on a master's degree, opening up a new range of professional possibilities.*" Quest Participant 2014.

Engaging with Nature during the Quests guides people to commune with the sacred in themselves and others; "*The Quest delivered everything I had hoped it would and much much more. I hoped to reconnect with a part of myself that has been absent for a while. I wanted to re-establish a meditation practice for myself, and I wanted to attune myself to a "higher" energetic frequency (for want of a better way of saying it). All of those things were achieved. What I did not expect was to get a real love for nature nor did I expect to experience a desire to immerse myself in it. I have been deeply afraid of the elements, particularly cold, rain and wind. I now see the possibility that I do not need to be afraid. The cliff walk to see the puffins was exhilarating. The solo was uplifting, enlightening and energizing. The energy work with the trees awakened something dormant in me.*" Quest Participant 2012.

Quest participants also discover that they can connect to Nature at a profound level, which opens them to experience and know the world and themselves in a new way. This is expressed by three of the people who attended my Quests as follows: "*When I "spoke" with the elder tree and it "spoke" back to me with such a force of impact I could not deny it even if I tried, this was an incredible affirmation for me of the existence of the energies that exist between everything on earth and in the universe.*" Quest Participant 2013.

«*I experienced profound shifts in my life and thinking by engaging with these activities" My heart opened and filled with the earth, the river and boulders, amazing trees, long vistas, birds, and their songs, flowers, and star-filled nights. Her special programs were powerful and reached the core of my being. Catriona's calm and clear energy quickly helped me shift to a grounded and revitalized state of being.*" Quest Participant, 2016.

Above all, Nature Quest allows people to experience themselves outside of the limiting construct of societal paradigms and the daily grind of activities that may not be in their best interests. The Nature Quest program is a pilgrimage to uncover the creative, authentic and wild heart within every person. "*My deep wish going into the Vision Quest was to experience myself in nature free of the distractions of daily life. I discovered a reservoir of inner strength, clarity, and fearlessness that I can now easily access in my day-to-day life. Catriona's attunement to each person's unique process created an atmosphere of trust and allowed me to surrender to each unfolding moment with an increasing sense of inner peace. There is no doubt that the Vision Quest is one of the most daring, rejuvenating and transformative practices I have ever done and I am profoundly grateful for the experience.*" Quest Participant, 2012.

«*I think the biggest benefit that I got was the chance to quiet my mind, ground myself, and get back in touch with my higher purpose in life. It was so good for me to get away from my regular routine, my "fishbowl" as I call the area I live in, and have the opportunity to learn new healing techniques. I loved the ah-ha moment I had when I learned "the time is NOW" to get back to my creative side. I have been writing and drawing every day since.*" Quest participant, 2015.

Thus, my knowledge of the powerful impact that spiritual pilgrimages into Nature can have on people's lives is not just based on research, scientific studies, or my own experience, but from witnessing firsthand hundreds of people reaping profound, life changing benefits. During my Nature Quests and Vision Quests, participants learn how to work with Nature to gain greater clarity and vitality and open doors to their creative potential. It is a way to allow the best that each person can be to spring forth, guided by the deep inner reservoir of personal truth that resides in all people. I see these immersive spiritual pilgrimages as partnering with and in Nature, where all the elements converge. In this way, people gain from directly interacting with each physical element found in Nature which then opens the door to the sacred dimensions of the Aether or Akasha.

~

Deepening our contact with the invisible "Akasha" will not only expand our worldview it will lead us to create and adopt better ways to live, including new ways of powering our physical world. This is already starting to happen as we abandon the concept of the "clock-work" universe. In the past, the aether was considered simply blank inert space, however today these outdated concepts are falling away.

While, Rene Descartes first introduced the concept of the aether in the seventeenth century to western minds, his discovery was ignored for hundreds of years. Descartes viewed aether as an active substrate that contained energy, force, and light.[13] He said that the aether was a fluid-like substance filled with vortices of movement and that this is why the planets move in an orbit around their suns. Over two hundred years after Descartes, Nikola Tesla went even further than Descartes to exclaim that *"All perceptible matter comes from a primary substance, or tenuity beyond conception, filling all space, the akasha or luminiferous ether, which is acted upon by the life-giving Prana or creative force, calling into existence, in never-ending cycles all things and phenomena."*[14]

Similarly, Paul LaViolette, a current-day astrophysicist, developed a unified field theory and refers to aether as the substrate from which all matter emerges.[15] LaViolette also believes in what he refers to as a "space vortex theory." According to this theory, matter arises from the aether as it is acted upon by cosmic influences.

Spiritual visionaries have shown that aether is not simply a passive "massless medium" but a "living fluid substrate" from which life arises. Science has caught up to this understanding, and in its wake unleashing alternative energy innovations, based on this new understanding. According to Paramahansa Tewari,[16] a former executive director of the nuclear power corporation at the Department of Atomic Energy in India: "*A century from now, it will be well known that: the vacuum of space which fills the universe is itself the real substratum of the universe; a vacuum in a circulating state becomes matter; the electron is the fundamental particle of matter and is a vortex of vacuum with a vacuum-less void at the center, and it is dynamically stable; the speed of light relative to vacuum is the maximum speed that nature has provided and is an inherent property of the vacuum; a vacuum is a subtle fluid unknown in material media; a vacuum is mass-less, continuous, non-viscous, and incompressible and is responsible for all the properties of matter; and that vacuum has always existed and will exist forever . . . Then scientists, engineers, and philosophers will bend their heads in shame knowing that modern science ignored the vacuum in our chase to discover reality for more than a century.*" To the modern physicist and engineer Paramahamsa Tewari, the ancient Vedic scripture *Idham thadhakshare parame vyoma* supports his theory that "the *aakaash*, akasha [or aether] is the primordial, absolute substratum that creates cosmic matter."[17]

The aether is not just filled with life-giving nourishment and powerful invisible forces from which matter arises. It also contains information, thoughts, and intentions. People who have attained a heightened spiritual alignment can work with this intelligent field of information. By working with this "information field" people can, and are, bringing new innovations into being.

For example, Paramahansa Tewari invented a "Reactionless Generator" (RLG) that achieves over 100 percent energy efficiency by pulling energy from the aether. Tewari's prototype was independently tested by Kirloskar Electric company in India and exhibited a 165 percent efficiency (over-unity), breaking known physics laws.[18] The RLG proves that thermodynamics first law, which states that a system's output can never be greater than the input, is not always applicable. By creating a machine with efficiencies greater than one, Tewari faced resistance from traditional thinkers. He proved that the RLG could run on a limitless energy source by harnessing energy from the aether without any harmful by-products.[19]

Tewari's RLG can be used for different applications, from lighting a room to powering a car. Since it can tap into free energy, its production and use would completely transform our world. No longer would people have to pay for electricity or gasoline. Unfortunately, without the type of extraordinary support that traditional industries like oil and gas receive, it is difficult to reach the public with these worthy inventions. Especially when existing industries seek to dominate the market and undercut inventions like the RGL that could provide free energy to the masses.

As we develop Earth-friendly technologies, we will be following in the footsteps of our ancestors, but with a new twist. We can learn to create and use energy in new and extraordinary ways to free humanity from economic servitude and level the playing field by providing free energy for all. Our descendants will use technologies like Tewari's RGL machine.

By working with, not against, the primal building blocks of the Universe, we are helping Creation unfold. Our inner world of thoughts, emotions, and ideas shape the outer world of experience and environment. As long as we connect to healthy, life-generating thoughts and actions, we can keep the wondrous dream of creation alive for our descendants. The mysterious aether and all its functions may one day be fully understood by humanity.

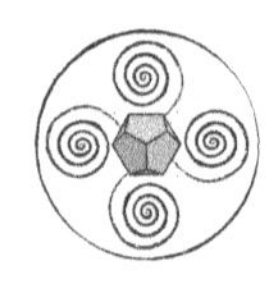

21. The Ring of Brodgar

A fo ben, bid bont.[1]

It was midnight, and I was standing alone under a full moon in the middle of the Ring of Brodgar, an ancient site in the Orkney archipelago. The air felt cool on my skin, and a soft mist clung to the ground. I come to the Orkney Archipelago every year in May to lead a Spiritual Nature Quest. Orkney is a place of contrasts. It alternates between past and present, bucolic beauty and stormy weather, intimacy, and grandeur. We hike to remote and beautiful places to view wildlife like puffins, harbor porpoises, and seals during the Quest. Participants also visit ancient sites and participate in a week-long spiritual program. One of the sites that we always visit is the Ring of Brodgar.

The Ring of Brodgar is one of the most magnificent and well-preserved megalithic stone circles in the world. Part of a UNESCO World Heritage Site, the Ring originally consisted of 60 standing stones, up to 15 feet tall, weighing as much as 40 tons, and surrounded by a deep henge.[2] The Ring is one part of a larger ancient complex. This includes the famous Stones of Stenness and the Ness of Brodgar. Nicknamed "Orkney's Neolithic Cathedral, The Ness of Brodgar is an ancient temple built in 3500 B.C.E.[3]

The Ness complex was surrounded by stone walls and contained huge free-standing buildings. The size and architecture of these structures that had painted walls and tiled roofs and close to a thousand examples of the decorated stone show that this was a special place in the Neolithic community. Today only a fraction of the original ancient structures have been unearthed, surrounded by thirteen-foot-high walls. Massive in scope, the entire multifaceted complex, which took several generations to create, was an important ceremonial and spiritual heart for the community.

The "Ness" was in use for over a thousand years and served the community much longer than any other cathedral. Many ancient sites in Orkney were built around the time of Sumerian brick temples and the simple mudbrick mastabas of Egypt's first and second dynasties. It dominated people's lives thousands of years ago, and it continues to dominate the landscape. The Ness builders were impeccable, as evidenced by the sandstone buildings and sites they left. Some buildings have walls 13 feet wide, with paved walkways, carved stonework, colored facades, and even slate roofs. These are as well-crafted, if not better crafted, than Egypt's structures.

Nick Card, from the Archeological Institute at the University of the Brodgar, discovered signs of magnificence in ruins: "*This is almost on the scale of some of the great classical sites in the Mediterranean, like the Acropolis in Greece, except these structures are 2,500 years older. Like the Acropolis, this was built to dominate the landscape to impress, awe, inspire, and perhaps even intimidate anyone who saw it. The people who built this thing had big ideas. They were out to make a statement.*"[4]

Simple yet distinctive pieces of grooved pottery thousands of years old have further proven that advanced civilization started in Orkney. The Orcadian temples and dwellings reveal the advanced architectural and artistic skills of the people who built them. Only later did stone henges like Stonehenge come into being on the continent and mainland Britain. Discovered by finding the same grooved pottery found at Stonehenge but from a much earlier time period. Digs at the Ness and elsewhere prove that the ancient inhabitants were advanced, not only in art and astronomy but also in agriculture. According to Jane Downes, the Archaeology Institute director at the University of the Highlands and Islands, "*Orkney's farmers were among the first in Europe to have deliberately manured their fields to improve their crops. Thousands of years later, medieval peasants were still benefiting from the work those Neolithic farmers put into the soil."*[5]

Orkney may seem remote today, yet it was an important central hub of the human world in Neolithic times. The oceans and the waterways were once the main highways with the Orkney archipelago at the center of major sailing and trade routes. Thus, while cities like New York, Paris, and London may be cultural hubs today, over 6,000 years ago, Orkney was a heart in the wheel of human society.

~

Standing barefooted and alone, with moonbeams illuminating a silvery mist, I was standing on a sacred bridge connecting the past to the present. The soft green grasses beneath my feet were still moist from the late-afternoon rain. The air was absolutely still but palpably expectant. From out of the dark, ephemeral faces began to appear. They were animated and alive, and yet ghostly as if existing in another realm and not fully present in this one. Orkney bestows the clarity that can be found in sacred places around the world. It's easy to receive visions of people and happenings from another time. On this night, I experienced a "download" that seemed to slip between the boundary uniting disparate worlds.

Faces spiraled before me as if they just escaped for a brief moment to send a message. Some of the larger and closer faces were recognizable: my parents and grandparents, and even great grandparents, I recognized from old photos. But as the faces receded further away on the spiral and were smaller and less distinct, it was impossible to place them. Was this the lineage of distant forebears making their presence known? Or friends and lovers, and possibly even enemies from a time long gone? Or a case of past, present, and even future relationships all existing together in this spiral of a continuum? These faces seemed far away and yet so near as if they were always beside me but hidden from sight.

The moonbeams shone on the massive megalithic stones encircling me. I thought of the people who carried these gigantic stones to build this amazing place that has survived thousands of years. I was humbled by the thought that I—their descendant—was now also standing by these stones so carefully placed ages ago. This humbled and amazed me. How in fact, could this ageless place still exist, in so much of its original glory, after all of the dramatic changes to the Earth, including wars, weather catastrophes, shifting human journeys, and disparate cultures claiming this land over the eons? Above all, in this place, at this moment, I am here with all those connected to me. We exist at once.

~

The Nature Quests I lead to Ancient Orkney reflect a familial and ancestral connection to Scotland's soil. While I was always aware of our MacGregor roots, and our mother taught us her Scottish sword dancing and instilled a Scottish sensibility in us, I only learned of our clan's past after my mother's death. It was indeed a miracle that a descendant of the MacGregor's, a clan that some sought to exterminate from the face of the Earth, would be standing here on this moonlit night.

On my first visit to Orkney, my sister Anne and I were traveling throughout Scotland over twelve years ago. It was a journey that brought us closer to each other and, unwittingly, closer to our ancestors. When we arrived in Stromness harbor by ferry from the Scottish mainland, we found ourselves in the midst of an ongoing international science festival. Scientists from over a dozen countries converged in Orkney to discuss their findings and feature their expertise. We decided to go to town later that evening to catch one of the many presentations being offered.

As we walked downtown after dinner, we stepped into a narrow alleyway with a cobblestone path across from Saint Magnus Cathedral. Each stone had been smoothed by the steps of many feet over hundreds of years. As we added our footsteps to those that came before and made our way along the alley to the slightly wider street, we came upon a small grouping of people waiting eagerly outside a church. Curious, we asked them what they were waiting for. They cheerily said they were going to see a talk by a Scottish DNA expert.

The group's enthusiasm was contagious, so we decided to join them. After waiting for just a few minutes, the church's thick wooden double doors swung open. A handsome young man with a reddened face, most likely a local farmer, smiled and said, "*Well come on, will ya before the* tirl *[gust of wind] catches you up.*" We all moved forward and entered and climbed-up the narrow staircase in the midst of the tight cluster of people chatting enthusiastically. When we reached the second story, we entered a large room filled with hundreds of seated guests. Somehow, we managed to find the last two empty seats in the Great Hall just in time for the presentation.

As we took our seats, the presentation was just beginning. The speaker introduced himself as Jim Wilson, Professor of Human Genetics in the Usher Institute for Population Health Sciences and Informatics at the University of Edinburgh. He explained that he would be discussing the results of his many years of studying DNA. He began his talk by explaining how the DNA study was conducted and the specific lineages his team had been tracking.

Astonishingly, midway in his presentation, a slide appeared with the bolded title "MacGregor" My sister and I looked at each other with amazement. He was about to embark on a genealogical study of the MacGregor clan. Not only was he studying our clan's DNA, but the synchronicity of the two of us being together in Orkney AND just happening to walk in on his presentation on a whim was remarkable. We were all ears.

Before delving into what we learned that day, it's best to provide a little background about the MacGregor clan, as its quite an unusual story. The MacGregor's are one of the most maligned people in history. In 1604, James VI (yes—the same James that vigorously hunted and killed" "witches") passed an Act of Council that outlawed the entire clan. According to the Act, *"The name MacGregor should be altogether abolished, and that all persons of that Clan should renounce their name and take some other name, and that they nor none of their posterity should call themselves Gregory or McGregor thereafter under pain of death."*[6]

Under this act, MacGregors' could not travel more than two at a time and were not allowed to bear arms or even carving knives. The proscription against the MacGregors was not permanently lifted until 1784.[7] The laws against them were not just patently unfair; they were downright cruel. So how did one people, one clan, come to be outlawed for almost 200 years?

Like most clans, the MacGregor's have a clan motto, and the MacGregor motto is "S' Rioghal Mo Dhream" which means "*Royal is my race.*"[8] Yet, the clan's reputation was under a cloud of controversy. In fact, just a few days previously, when Anne and I were camping on Clachtoll Beach on the West coast of northern Scotland, I was unceremoniously reminded of this fact. I had struck up a polite conversation with an older Scottish couple. After conversing pleasantly for 20 minutes or so, I mentioned that Anne and I were of the MacGregor clan. Almost as soon as I had finished saying the name, MacGregor, the husband appeared to rise a bit taller and said,"Och, you MacGregors were Baaad." Half hoping he was joking, I did not respond, but our conversation ended abruptly as he turned and left the building without saying goodbye.

While some claim that the clan's descent from kings was pure mythology, there is a lot of historical information that says otherwise. For example, in the 1850 book Clans of the Highlands of Scotland, Silbert Thomas declared that: "*All men admit the clan Gregor to be the purest branch of the ancient race of Scotland now in existence—true descendants, in short, of the native stock of the country, and unmixed by blood with immigrants either of their own or of any other race. About this point there is no dispute; and the name of Clan Alpine, commonly adopted by them for centuries, would almost alone suffice to prove their descent from the Albiones, the first known inhabitants of Scotland.*"[9]

As my sister Anne and I sat on simple wooden folding chairs in the back of the church auditorium, Jim Wilson interpreted the hundreds of MacGregor DNA samples he had obtained. He explained that he expected the DNA to be similar to most modern-day Scots at the outset. Yet, over time as results came in, he found that MacGregor DNA had a different chromosomal marker. The chromosome belonged to those of ancient Pictish lineage and specifically to the Dalraidic kings of old. Thus, his findings affirmed oral history and older literary references that the MacGregors are indeed descendants of King Kenneth MacAlpin, the first King of Scotland.

Kenneth's mother was Pictish, and he likely came to his kingship via matrilineal succession. Kenneth took his throne as both "King of the Pict" and "King of the Gael" via ancient ritual by sitting upon the "Stone of Scone" on the land.[10] This stone, also referred to as the "Stone of Destiny," is made of sandstone, with a hollowed-out surface for a chair.

On this stone, all new kings were coronated out of doors to seal the new king's relationship to the land he was to rule.

Kenneth's uniting of Alba (Scotland's first name) was the beginning of the fusion of two peoples who already shared many similar customs and beliefs. The uniting of the Picts and the Gaels was driven in part by a need to defend themselves against the Romans and Viking raiders.[11] Yet, it was not just the Romans and the Vikings from the North who gave the Picts and the Scots a hard time.

In 1066 the Normans conquered England. They later pushed into Scotland, winning feudal ownership of the island in 1072.[12] These invasions were led by Duke William II of Normandy, later referred to as William the Conqueror. He was a direct descendant of the Viking warrior-king Rollo, who led Normandy's invasion in 918 C.E., William became the first Norman King of England, reigning from 1066 until he died in 1087.

Lands formerly held by Pictish and Gaelic families and clans were seized and later deeded to Breton, French and Norman soldiers, and nobles loyal to the new king. In 1292 King Edward of England commanded that over 1,800 Scottish nobles pay formal homage to him or die.[13] Norman and Anglo-Saxon Kings seized the Scottish throne, often betraying or killing other potential claimants who had ancient roots to the land.

Over time, the more ancient clans with a lineage going back to the royal line of the first Kings of Pictland and later Scotland, like the clans MacGregor, Grant, Macaulay, Mackinnon, Macnab, Macfie, and MacQuarrie, lost the majority of their lands.[14] It was given to clans like the Campbells and Stewarts, who fought for the Normans. Once MacGregor land was taken away, the new landowners would sometimes burn the dwellings to the ground, keeping the statelier castles and buildings for themselves.

By the late 1500s, some of the most beautiful lands in Scotland, held for generations by the MacGregors, were taken from them until they scarcely owned a single parcel. These seized lands included Loch Awe, Glen Orchy, Glen Strae, Glenlyon, Strathfillan, Balquhidder, and Breadalbane, including lands near the Loch Katrine and Loch Lomond, stretching all the way to the sea. After being pushed to the brink of ruin, the MacGregors eked a living by hunting game on land now owned by others and practicing cattle thieving. In those days, one's wealth was in land and cattle, and without either, one had to become a tenant on someone else's property or poach wildlife and cattle to survive. Yet, not all MacGregors buckled under or were killed.

Rob Roy MacGregor was born near Loch Katrine in 1671 to Donald MacGregor, a senior clan member.[15] Rob Roy lived during the terrible proscription against the MacGregors. His father, Donald, was imprisoned for the Jacobite uprising and killed. Rob Roy was considered a folk hero among the local inhabitants. He later became romantically likened to a Robin Hood-type character by outfoxing cruel lairds (Gaelic for lords).[16] Rob was well-educated, fluent in two languages, Gaelic and English, and was an excellent swordsman.[17] He also had a good sense of humor, as shown by his letters about his turbulent relationship with the Duke of Montrose. Rob was also very beloved by the local community, yet his life was often in danger.

The Duke of Montrose, the son of a Breton Knight and mercenary, who assisted William the Conqueror in subjugating Scotland, chased Rob Roy for almost ten years. Yet, Rob Roy always managed to outsmart him. When the authorities came hunting for Rob, he would often hide in small caves along their banks. Some of these had entrances that appeared too small to hold a man of Rob's size but expanded into larger spaces further back. These naturally occurring hiding places preserved Rob Roy's life on many occasions. Rob Roy was not the only one who depended upon natural hiding places. Some MacGregors managed to live by moving further up into the wilderness areas of the highlands and becoming astute at slipping away from the sight of man.

They became known as "*The Children of the Mists.*"[18] " *Children of the Mist*" reflects their ability to synchronize their movements with local weather patterns, disappearing into the thick Scottish mist, which often covered the mountains and lowland areas. Ironically, the unjust proscriptions against the MacGregors created a clan known for its independence, stealth, intelligence, strength, love of Nature, and, above all, loyalty. Perhaps the "*Children of the Mist*" had a hand in shaping my destiny to be a lover of wild places. Their ways certainly led to their survival and descendants that outlived the unfair laws and lived freely, using their names. It's not just the MacGregors who have seemingly miraculously survived, but the Picts themselves.

While some consider the Picts an extinct race, they are still here living through their descendants and beliefs, and practices passed down through generations. When the Picts and Gaels united under Alba, they united as a people, intermarried, and adopted each other's customs and ways. Thus, Pictish influence goes beyond just genetic influence, as it also affected Celtic and Scottish culture. According to historian Stuart McHardy, the ancient Picts certainly "*did not disappear*" given that their descendants still "*walk their lands and fields in the present day*" Moreover, McHardy points out that the Picts had a major influence on Scotland's clan system development, along with "*many of the cultural, artistic and philosophical leanings of Scotland's people.*"[19]

MacGregors survived despite incredible hardships. Yet, there is scarcely a lineage of people on Earth who have not, at some point, been treated unjustly or faced hardships in some way. Their singular story reflects the indomitable human spirit's ability to adapt, survive, and thrive.

In fact, the entire human race almost became extinct in 70000 B.C.. Geneticists have indicated that there may have been as few as 40 breeding pairs of humans.[20] If there had been an Endangered Species Act back then, humans would have been on the list. Many scientists have linked this near-extinction event to the catastrophic eruption of Mount Toba. Toba spewed 650 miles of vaporized rock and gas into the atmosphere, causing the sun to dim for over five years.[21] While archeologists are conflicted about the degree of its influence on the human species, it's clear that our numbers declined to a worrisome low based on DNA evidence.

Humans worldwide should feel a familial sense of relationship since DNA now proves that everyone on Earth is related to everyone else. Every human being who was alive 10,000 years ago is the shared ancestor of everyone today.[22] If you go back 500 generations, all of us have a common relative. If you go back just 100 generations, or 5,000 years, when the ancient Pictish people were still using the Ness of Brodgar and living in their Brochs, a family reunion would require almost everyone on Earth to be invited as Kin.

This gives one pause when considering the scope of one's relations. Maybe on that chilly, moonlit night standing alone in the Ring of Brodgar, I glimpsed faces of a common ancestor from hundreds or thousands of years ago. An ancestor who wishes to warn us to change our ways for the benefit of humankind or let us know that ultimately, we will be victorious in ensuring the human race's continuation. One thing is for certain; they would have considered the scope of our relations to go far beyond our singular human species and encompass all life on Earth.

~

In the spring of 2015, I arrived with a new group of participants for the Scotland Quest. I invited Nick Card, the Ness lead archeologist, to join us for an informal fireside chat to present his archeological findings of the Ness of Brodgar. Sitting by the fire in the cozy parlor, a welcome treat after our exhilarating hike earlier to windy cliffs to see puffins, Nick brought a replica of a carved Pictish ball identical to the original in size, weight, and design.

As Nick carefully removed the sphere from its box, we were transfixed by its compelling shape and thoughtful design. We took turns holding the sphere as Nick described how the original had been found in a very prominent place within the ancient cathedral.

These perfectly shaped balls reflect the ancients' interest in and familiarity with sacred geometric designs. Many of their engravings, jewelry, and art repeat natural patterns like the spiral. As the beautiful sphere passed from hand to warm hand, those holding the stone fell silent as if mesmerized.

As the stone sphere made its way around our circle, and I finally held it's smooth and now warmed shape in my hands, I felt a deep connection to the ancient people of Orkney and my own ancestors. Once upon a distant time, they sat holding a stone sphere, with orange flames and visions dancing in their eyes.

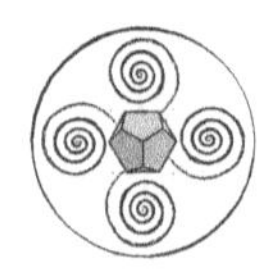

22. Cloud Computing

The development of full artificial intelligence could spell the end of the human race.[1]

Stephen Hawking

Almost two decades after my flight above the Laguna Madre, I attended the Digital Ecosystem Conference in Silicon Valley, California. I had been invited by a friend who worked in the IT business, and he thought it would be interesting for me to learn about this innovative new software. We sat at the back of the room, at a table covered with a white tablecloth, with six others. Sitting in a large conference room filled with over two hundred attendees, we both were seated at a large circular table enjoying lunch. Although the conference had the word "ecosystem" in its title, I knew we were not about to hear about living organisms.

Standing in front of an enormous screen at the front of the room, the young presenter made a very engaging speaker. The exuberant twenty-something man, in faded jeans and sneakers, with a wide boyish grin, told us that he and his team had launched something that was a "*tremendous advancement for humankind that would change the world as we now know it.*" My friend turned to me and smiled. I was less enthused but certainly on board to learn what I could.

When the PowerPoint powered up, thirteen circles in bold red, blue, green, and yellow floated on a white screen. As he spoke, the presenter added to the diagram. Dotted lines connected some circles, and then other circles were connected by straight lines. The circles represented different companies and computer stations, and the lines indicated how they were to be connected. With considerable flair, the presenter clicked his controller, and one large white oval appeared at the very top above all of the circles and lines. Inside the large white oval was the word "CLOUD," and above that, in bold lettering, were the words "Digital Ecosystem." The presenter then positioned his light pointer on one of the colored circles and exultantly said, *"Every computer can now be connected to the cloud and every other computer. They are all now in the same 'ecosystem'."* His statement was delivered with as much bravado as if he had just invented a new species, and in a way, he had—as far as human technology was concerned.

As I looked out across the conference room at the hundreds of people gazing intently at the screen and smiling, a depressing thought flowed into my mind. I was probably the only one in the room who was uncomfortable referring to a network of machines as if they were a living ecosystem.

Do not misunderstand. I do believe that the internet and computers are amazing tools. I use them quite a bit myself. Yet, the problems with this technology are many and include; addiction, harmful radiation, replacement of time spent socially with friends and family, lack of privacy, AI taking over people's lives, and more.[2] These serious problems are not being tackled with the same enthusiasm as expanding and "improving" the technology. Hearing this new software being hailed as, well, a living thing, riled me.

"Ecosystem" refers to a living system comprising a community of life forms and their habitat. Scientists and environmentalists have been using the term to describe how the plants, the animals, and even the air and soil are connected in life-generating ways. A true ecosystem is, in essence, a complex whole that depends upon all of its parts. Take away even just one of these parts or add something new, and anything from a quiver to a shock wave will move throughout. Take away too many parts, and the entire ecosystem may even collapse. This may be a well-understood concept today, but it was very hard-won.

For hundreds of years, scientists studied single species as if they existed in their own separate universe. This kind of objective thinking led to demonstrable harm as we have exterminated certain species from their ecosystems, transplanted non-native species, and damaged habitats beyond repair. This "life isolationist" theory that proposes that each life and species exists in its own bubble still, unfortunately, drives scientific experiments and leads to the creation of harmful "products" like genetically modified organisms. By usurping Nature's language and applying the term "ecosystem" to refer to machines, we dilute knowledge recently gained.

It's taken many generations for humans to shift from a mechanical and isolationist world view. We may take little notice that "ecosystem" is now an active part of our vocabulary, yet it reflects many years of research and education. This is why the term "ecosystem," to a scientist who studies life, or an environmentalist who seeks to protect that life, is sacrosanct. The term itself reflects and represents an entire paradigm of an interconnected living Earth. Thus, the use of the term "ecosystem" to describe an assortment of artificial electric boxes and their wires on that day in Silicon Valley felt— well, wrong.

That is not to say that there are not many things I appreciate about the internet and computers. First, the internet is one of the few truly democratic communication mediums today (although moneyed interests and government control are now trying to take it over). It has sparked an information age like no other, bringing news to people worldwide and helping them connect. When I was the Executive Director of an international environmental NGO that supported a large network of environmentalists in China, Russia, and the Pacific Rim, we could discuss complex solutions to global problems easily because we could communicate across vast distances in real-time. Yet, as with all completely artificial technology, misuse or abuse can lead to harm. Our society's over-reliance on the internet and other man-made technologies is indeed causing harm.

It is estimated that one in eight Americans suffers from Internet Addiction Disorder. Addiction is considered the correct diagnosis when indulgent use takes over people's lives. Up to 30 percent of South Koreans under 18 have this disorder.[3] Ten percent of China's teenagers suffer from internet addiction; Chinese officials who run internet addiction centers refer to it as "*electronic heroin*."[4] They believe it as being just as challenging for an addict to break away from the internet as heroin. However, unlike heroin, the internet is free or relatively inexpensive. Without any cost barriers to its use, it is spreading like wildfire, especially among teenagers. It especially concerns me when young people overdo it. Demonstrable shifts occur in the brains of people suffering from internet addiction, especially those that start using it at a young age as the brain is developing. These shifts leave the addicted person with an inability to control their attention or make executive decisions.[5] It also impedes their emotional processing making it harder for them to read human emotions and socialize effectively.

Our reliance upon a computerized and technological world has gone too far. We are replacing the real world with phantasms on the screen. We are becoming a society out of touch with reality, which leaves us blind, deaf, and dumb to what is happening just outside our door. Not only are many out of touch with Nature, but some also find the illusory world on the screen more stimulating and attractive because they can apply optimum control to what they choose to experience. This effectively erases the natural world's influence on one's experience and understanding because no one pays enough attention to the natural world.

Technology monopolizes people's time and focuses them on a synthetic world. While technology is super useful when it serves us, technology can limit our species' potential when it takes over our lives. We may even be altering our evolutionary path by changing our brains. By turning away from the natural world to see the glowing screens of an artificial one we disconnect physically and psychically from the real world.

We know that the technologies we rely on most, like cell phones and computers, use artificial frequencies. We also know that our state of spiritual awareness is directly linked to measurable "frequencies." Certain frequencies tend to be more conducive to meditation, hands-on healing, and spiritual enlightenment. When we create an artificial "zone" of frequencies in and around the entire planet, these can mask natural biological rhythms that connect us to the Earth, Nature, and our own Souls.

Luncheon plates clacking together loudly brought my attention back to the Digital Ecosystem Conference. Efficient serving staff in black jackets and white shirts cleared our tables and whisked quickly in front of the final slide that loomed large on the screen. The background image was of a stark blue sky with one giant cumulous cloud. In the foreground, large bold letters strung together two words that normally, or should I say naturally, never fit together: CLOUD COMPUTING.

The slide brought to mind a recent photograph I'd seen of a rather surreal-looking cloud that shimmered in the early dawn sky. I had nowhere to place it in my memory. The photo was of noctilucent, or "night-shining," clouds. These clouds are a new phenomena that, as far as we know, appeared at the turn of the twentieth century as the industrial era took hold.

Noctilucent clouds are found in places where ordinary clouds do not belong. Floating, or rather spreading, fifty miles high in the Mesosphere, noctilucent clouds are the highest clouds around the Earth. They are caused by the combination of water, ice, and dust particles and can only be seen (very faintly) when the sun is just below the horizon. Noctilucent clouds appear to be occurring more frequently, and they have been linked to pollution and human activity.[6] These clouds are not like the puffy white cumulus clouds that conjure up pleasant floating castles in a blue sky. It was as if the Creator had accidentally spilled milk high above the Earth and absent-mindedly left it dimming the light from the rising sun. I wondered if anyone at the conference knew about them?

Perhaps these "clouds" are somehow related to the large hole in our Earth's stratosphere. The stratosphere is the Earth's middle and in some ways most important atmosphere of the three layers that surround the planet. The stratosphere protects the Earth from too much and harmful UV radiation from the sun. Without the stratosphere, which acts a shield to stop almost 99 % of harmful radiation, there would be little or no life on Earth. However, in the 1970's it was discovered that the stratosphere was thinning significantly, and a "hole" opened up above Antartica. The cause was determined to be chloroflurocarbons (CFCs) and halons used in industrial production of air conditioners and refrigerators and other products. These products were banned under the Montreal Protocol on Substances that Deplete the Ozone Layer in 1987 and phase out began in 1990.[7]

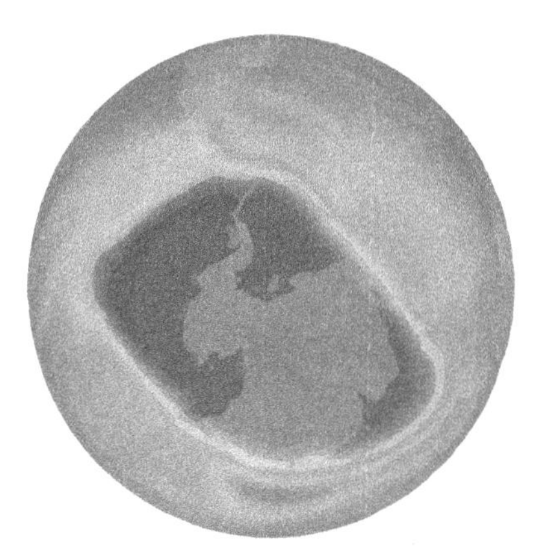

Ozone Hole, September 16, 2020

Despite these efforts, the ozone hole swelled to almost 28 million square kilometers in 1998 and covered 25 million square kilometers in 2020. (Note that the hole is seasonal and tends to be largest in the months September and October.) This is because the CFCs and other harmful chemicals take many years to flush out of the system. In addition, the newest 2018 "Scientific Assessment of Ozone," has found that emissions from eastern Asia have increased, and that other harmful chemicals are still affecting the stratosphere.

Increased UV radiation, especially the increase of UVB radiation is having demonstrable impacts on planetary life. For example, there are significant more skin cancers and cataracts due to the reduction in protective atmospheric ozone. It has even been found that too much UVB can harm the human immune system. We are also seeing a great deal of impacts to plant life. UVB can sterilize trees, and cause sunscald on their leaves, making it more difficult to conduct photosynthesis. Damage to plant DNA is also occurring and can inhibit leaf expansion and/or modify normal growth patterns. Damage to Earth's atmosphere is one of the most devastating things that mankind has done.

As the conference ended and the room began to clear out, a handful of people stood around the presenter to ask questions or introduce themselves. I sat for a moment alone at the table after everyone, including my friend, had left. As I placed my napkin down, I knew that while many people will eventually hear of and perhaps even use cloud computing, few will notice the unusual appearance of synthetic clouds nor miss the clear blue skies of yesterday. Even fewer know that we have opened a hole in our Earth's protective atmospheric shield. As I left the building and drove home, I could not shake the uncomfortable feeling that had overtaken me. What if future generations become so enthralled by technology and AI that they spend most of their time in a virtual world? What is worse than allowing the world to be destroyed? The fact that no one would look up from their screens to notice it was gone until it was too late.

We must take care that computers do not cause us to replace a three-dimensional experience with a flat-screen. Throughout the book, I have shared how important it is to experience animals, trees, and the Earth directly. We must not become overly enamored with machines that can only offer a flat, two-dimensional world. No technology on the planet can replace being loved, cared for, or understood by another living being.

Lying just beneath the conversation about computerized ecosystems is, I believe, unconscious neglect, or perhaps even an egotistical disdain that humans are smarter than the natural world. If we replace Nature in our lives with a fascination with a fabricated emotionless technology, we face a world that is impoverished of biodiversity, and may lead to the end of us.

VI. ADAPTATION

23. Younger Brothers & Sisters

We live within a self-regulating, interconnected, multidimensional system that wants life to succeed, and is continually creating more life.[1]

CATRIONA MACGREGOR

The Kogi live gently on the Earth, exhibiting a vast understanding of ecological principles, human nature, and even the stars. They gained this knowledge not by dissecting the external world as if it were separate from themselves or using harmful technology, but by accessing the most sacred of their human capacities. Kogi believe that human beings can change Nature in extraordinary ways. By doing so, we can benefit humanity and all life, but only if we first understand and then work in partnership with the elements, the plants, the animals, and Aluna herself.

Indigenous peoples like the Kogi tend to share key beliefs in common: first, they have a profound understanding of the natural world and natural processes; second, they exert only minimal control on Nature, perceiving it as a whole and good in and of itself; third, they have a rich spiritual life that includes a deep connection to the Earth and the invisible forces that support our existence; and, fourth, they honor and practice reciprocity, and understand the need to give as well as take.

According to the Kogi, we, younger brothers, "*can learn to see the changes, too.*" "*Open your eyes and awaken,*"[2] they say. When we see the world the way the Kogi do, we will treat all living things justly. By adopting a new societal paradigm; that we are here to give more than to take we finally become the "Wise Humans," and the "Real People." First, though, we must make the changes; we must adapt. The process of adaptation begins as we shift our perception and world views. Renowned Chinese marine scientist, Dr. Wang Pelie, is one of the many people I have met who dramatically changed their world view in an instant.

In the 1990s, I launched a China conservation program and hired an in-country representative and program leader named Wen Bo. Wen Bo is an exceptional leader of the growing environmental movement in China. He founded the China Green Student Forum, a network of more than 100 student environmental groups. Wen was also an Asia 21 Fellow of the Asia Society, received a 2009 Pew Fellowship in Marine Conservation, and was named a 2009 Young Global Leader by the World Economic Forum. By building a dedicated network of college students, scientists, and agency officials committed to similar goals, Wen has moved China's conservation movement forward in no small way. Time Magazine interviewed Wen and premiered him as one of the country's eco-leaders.[3]

Wen helped to oversee our China environmental program, where we supported over forty environmental grassroots groups and scientists. He suggested I focus on supporting marine conservation, which tended to receive less attention and support than terrestrial conservation programs in China. We hosted a marine conference in China's northernmost city of Dalian to support marine scientists and conservation groups working on marine research, restoration, and protection. One of these scientists was Dr. Wang Pelie.

On one of my visits to China, I headed far north to Dalian, one of China's deepest water ports to meet with scientists and prepare for our upcoming marine conference. In the morning, Dr. Wang Pelie, met us on the steps of the aquarium. A lean and unassuming gray-haired man in his seventies, Wang spoke quietly and suggested he would give me a tour inside the cavernous building. Just inside the entrance was a giant, bleached-white whale skeleton hanging from the high ceiling.

As we stood beneath it, Wang spoke knowingly of the whale's migration routes and where this one was caught. As we continued our tour and conversation, he explained two of his marine research projects in detail. One project was to study manatees, rare, warm water-loving mammals found near Hainan Island, a semi-tropical habitat in southern China. Historically, these unusual animals have been mistaken for mermaids by sea-weary sailors. Their species had declined rapidly, and he was seeking a way to learn more about them to protect them from becoming further endangered.

He was also seeking support for his team's research on the spotted seal. Spotted seals have lost important breeding grounds on the coast due to rapid climate change. By Dalian, the winters can get exceedingly cold in the north, with massive ice flows off the coast. The spotted seal depends upon these icy havens for its young. With global warming and melting ice, predators could reach the seals and eat the unprotected babies. Thus, their population was dropping precipitously.

At the farthest reaches of Chinese northern and southern borders, these two mammal species were struggling to survive. Wang's work was one of the steps needed to ensure their protection and recovery.

Nearing the end of our time together, and as we walked along a long hallway between exhibits, he slowed his step as if deep in thought and then stopped. Leaning slightly back on the banister, he began to speak in a thoughtful and heartfelt manner. He said he wanted to tell me how he first came to protect marine mammals. He told me that he worked for a large whaling company for the first 25 years of his career. With his knowledge of whale biology and behavior, his job was to find whales to kill for the market. In an almost confessional tone, he revealed that he had helped to kill many whales, including the one whose white skeleton we had stood under earlier. Sadness filled my heart at the thought of all of those beautiful, wise behemoths cruelly felled.

Wang explained that he was usually below deck when the whales were actually harpooned. "*Then one day,*" he said," *once we were far out in the ocean, I went up on deck as the ship approached the next whale victim. The lethal spear shot out from the ship's bow, and the great whale groaned and spun in agony. The whale's death throes brought it close to the side of the ship, where I stood. As I looked down into the water towards the great dying beast, I saw the whale's huge eye looking directly at me as it died. I was stunned. I saw for the first time that the whale was a conscious being. It had a soul. "What have I done?" I thought to myself."*

The instant he related to the dying whale, his perception shifted dramatically and instantaneously. He felt the agony this great being suffered and his perception about whales being merely products for the market changed profoundly and forever. After experiencing deep despair for killing these magnificent beings, Wang soon left the whaling company. For the next 30 years of his life, he has devoted himself to educating people about the importance of whales and other marine life, conducting research, and advocating for their protection.

One day, all people will understand that all life has a degree of consciousness. A whale, a bird, or a deer may have a different quality of consciousness than us, yet they all carry the same sacred spark from the Creator of all things. The beauty of Dr. Wang's story's is that the change of heart that he experienced is a possibility for everyone. This is why, even though I am an ardent environmentalist, while I may hate the actions of those who abuse animals and destroy Nature, I try not to hate the person. I always hold out hope that someday they will change. Just as the whale's soul spoke to Wang's heart, the soul of Nature can speak to and redeem the human heart.

24. The Vision

Would the valleys were your streets, and the green paths your alleys, that you might seek one another through vineyards, and come with the fragrance of the earth in your garments.[1]

Kahlil Gibran

The song of the Orkney skylark resounded in every blade of grass, shimmering with drops of dew filled with dawn's yellow light. As the skylark climbed higher and higher it disappeared into the sky and dropped its song as if it were a gift from heaven. I finally pinpointed the songster, a male skylark in full display, lowering himself whimsically from on high with fluttering wings. The bird hovered over one of the largest standing stones as if it was the stone's feathered crown. Then rose 300 feet into the air, filling all the land with his sky-song.

Birdsongs can contain complex notes and beautiful rhythms, and the skylark has one of the most complex and fascinating songs of all. Each male bird has his own special song, and some of them sing over 300 different syllables like pearly notes on a celestial necklace. Skylarks tend to sing the most at dawn as if celebrating the Earth's awakening. This morning, the skylark's song resonated deep within my being. It felt as if the very cells of my body were smiling from the lyrical sound.

Birdsong is not only beautiful to hear, it can also be healing. A scientific study in England showed that birdsong helped hospital patients heal more quickly.[2] Birdsong reduced patients' stress and pain response and improved their mental health. People that are surrounded by bird song also tend to experience improved mental health and greater moments of happiness. This was proven by a study launched by the University of Exeter, the British Trust for Ornithology, and the University of Queensland that measured the mental health of over 300 people.[3]

Birdsong can even improve the well being of plants. According to a recent study published in Scientific Reports, researchers from Yeungnam University in South Korea found that birdsongs cause plants to undergo positive biological transformations such as increased growth.[4] These natural sounds also enhance the plant's size and total mass and the volume of seeds and crops it produces.

Hearing birdsongs is an enhancing experience for humans. We certainly know it's psychologically up-lifting. Saint Brigid was said to have woken to her prayers into the sound of the lark. In mythology, the bird is linked to daybreak, new beginnings, freedom, and joy. Percy Bysshe Shelley captures these sentiments in his poem "To a Skylark":

Hail to thee, Blithe Spirit
Bird thou never wert,
That from heaven, or near it,
Pourest thy full heart
In profuse strains of unpremeditated art.
Higher still and higher
From the earth thou springest
Like a cloud of fire;
The blue deep thou wingest,
And singing still dost soar, and soaring ever singest.
In the golden lightning
Of the sunken sun,
O'er which clouds are bright' ning,
Thou dost float and run,
Like an unbodied joy whose race is just begun.
Better than all measures

Of delightful sound,
Better than all treasures
That in books are found,
Thy skill to poet were, thou scorner of the ground!
Teach me half the gladness
That thy brain must know,
Such harmonious madness
From my lips would flow
The world should listen then, as I am listening now.[5]

As the skylark's joyful notes rang out that morning over the fields, reverberating against the ancient stones, I had a stunning revelation. The vision from the Spirit Horse did not end at the shocking transformational touch of the cosmic ribbon. Stirred by the skylark's song, the remaining pary of the prophetic vision burst forth from my subconscious. While the Spirit Horse and I seemingly vanished into thin air, I learned that this was not the end but just the beginning of the Vision and a dream of a possible future for humankind.

After the cosmic event, I discover that I am flying or floating just above the Earth's surface, at the altitude of a single-engine plane in cross-country flight. Beneath me, an expansive forest of healthy dark-green trees stretches as far as the eye can see. At the edge of the forest stands a tall mountain. Two-thirds of its massive sides are covered with trees and shrubs, but the last third, the mountain top, looks like solid granite. I fly towards the mighty mountain and circle around its peak. A small village sits at the base. As my focus turns towards the village, I am instantly transported within the village itself, hovering just a few feet above the ground.

The village at the foot of the mountain consists of modestly sized white "buildings" made of smooth material with a lustrous sheen. The humble-looking buildings are organically shaped. Rounded and curved walls flowing seamlessly into roofs and floors. These beautiful and natural-looking dwellings have a wholesome feel about them, as if they were "grown," not constructed. I reach out and place my hand on one of the building walls, and the material feels cool to the touch and oddly vibrant. These exquisite buildings remind me of large, beautiful seashells. The seamless appearance leads me to believe that they were created through natural processes. It feels good to be in their presence, and a sense of innate harmony fills the homes. I imagine that residing in these homes would be incredibly healing and peaceful.

The buildings are clustered close to each other, 20 to 40 feet apart, with walkways three to four feet wide between them. I get a sense that this is a very close-knit community. The sounds of birds and animals from the nearby forest can be heard within the peace-filled village. There are no loud motorized noises to drown out or distract. The air is clean and aromatic, with gentle breezes carrying pleasing perfumes from nearby flowers and plants.

As I move around the village, I follow walkways that meander gracefully throughout. There are three walkways wider than the rest that connect individual homes. One wide walkway leaves the village and heads towards the unending forest. I understand as if tapping into an innate knowing about this village and its inhabitants that the residents spend a good deal of time in the forest to study Nature. This is where the children learn about the animals and the plants and the workings of natural processes.

Another of the three-wide walkways leads to a large open field with many different kinds of healthy-looking crops. This is where the inhabitants, who are vegan, communally grow their food, and contribute to the garden. The plants are treated to pure, clean water and special nutrients to create a "high-vibratory food source." The plants look extremely healthy and vibrant; many are unusually large. All types of fruits and vegetables are being grown, some are familiar and others I have never seen before. Some trees at the edges of lower vegetable gardens produce several different varieties of fruits and nuts.

The last of the three-wide walkways lead to a gleaming white building. While the building appears seamless and similar to the homes, it is much larger. Besides, there are lovely designs of animals emblazoned into the surface, geometric shapes, and unusual looking lettering. This wondrous looking building radiates strong energy that makes me feel instantly uplifted. I wonder if the markings are a symbolic language or if the geometric shapes somehow enhance the structure and surroundings.

I understand intuitively that they are doing both. The entire building has holographic qualities made from combining beautiful frequencies. The symbols are an ancient cosmic language that has been rediscovered and is now used as a universal language throughout the Galaxy. The shapes reflect harmonious cosmic interference patterns produced by life-generating frequencies.

The building is a sacred community gathering place, and the residents understand how to use shapes and sounds to heal and uplift their community and the Earth. Their beliefs and understanding about the sacred are much more defined and realized,' as well as their commitment to supporting not just their own personal wellbeing but the wellbeing of their community and the planet.

I spend some time just outside of the entrance, absorbing the inspiring energy, and looking closely at the geometric forms. A few are very familiar and remind me of the platonic solids. One looks like the Merkabah, another like the flower of life. Another looks like a replica of the green geometric form that appeared to me after my year-long water fast.

My curiosity gets the better of me, and as soon as I wonder why there are no roads, I am instantly transported away from the Temple to just above the mountain top. Except this time, I am not by the village but on another side of the mountain. At this close range, I can see that the mountain is too smooth and perfect to be a natural feature. The mountaintop is of a different material molded to blend seamlessly with the natural granite base below.

Flying in a circle around the top, I stop suddenly at the peak just above the village. Two enormous clear windows are set into the peak. These monumental "windows," at least 1,000 stories tall or taller, are tilted upwards at a slight angle. I know that they cannot be made of glass, as glass can't hold that size. Yet, the windows are clear like glass and provide an optimum view of the sky and beyond. Behind the "windows" and inside of the mountain top itself is a cavernous space.

The interior is lit within by an opaque white light that emanates from the pale gray walls and ceiling. There is one large lower floor with several layers of half floors that reach towards the ceiling. Sitting on the bluish-gray floors are hundreds of unusual but beautiful vehicles of flight. They range in color from pure white to off white and light grey and black. Their surfaces look smooth but not shiny. They bask in the light and do not reflect it.

Some of the craft are triangular-shaped or manta ray shaped, others are rounded and pod-like; some are long and cigar-shaped. Two of the craft are large enough to fit thousands of people; many others look like they make for a comfortable ride for four to six. Smaller craft appear capable of flight in the Earth's atmosphere, while others lack aerodynamic features and are more suitable for flight beyond the Earth's atmosphere and into deep space.

I understand that the mountaintop is a sophisticated space station and launchpad for spacecraft and aerial vehicles. Residents not only study the galaxy and the cosmos in stunning detail and with great accuracy from here, but they can also travel out into the solar system and beyond. Now I understand why there are no roads in the village; people move freely all around the planet in these energy-efficient, nonpolluting craft.

While the community that lives in the village appears low-tech, with simple dwellings, few possessions, and little signs of physical changes to the land, it's clear now that they have technology far superior to ours. Beneath the stargazing mountain, inhabitants use these amazing "machines" to fly to different destinations on the Earth or even out into the cosmos.

The vision of these unusual and beautiful flying machines is exhilarating and oddly familiar. I am drawn to them. I am instantly inside one of the smaller craft, admiring its interior's sleek and simplistic beauty. A central column encloses the "power" source for flight, and on one side of the craft is a smooth white panel with a few raised buttons. Their synchronous movements match the pilots' brainwaves and intentions.

These are "living" machines that have a form of intelligence. I look up above the panel, and the wall before me, the floor below, and ceiling above turns first opaque and then clear. I can now see out of the craft towards the enormous windows. Outside and beyond the windows, the sky's blueness fades, like a veil opening, to reveal stars and planets beaming from the darkness of space. I face the very center of the Milky Way Galaxy. As I long to explore those mysterious depths of space, I feel the craft lifting and moving towards those brilliant suns' and lights to other worlds.

The lights then begin to move gently back and forth, like a soft wave flowing this way and that. Soon, the lights begin to look more like flowers with defined shapes and colors. I find myself now back again riding the Dream horse, running in a field filled with grasses and flowers.

~

This prophetic Vision lifted the great sadness that had burdened my heart for many years. I had been mourning for the loss of so many for so long. So many trees, animals, and species have disappeared from the Earth. Yet, there was a hopeful and inspiring vision. A Vision that showed that humanity could live in harmony with the Earth and all life.

Now that this new world lives inside of me, I offer this Vision, this prophecy, as a song-line cast faithfully forward into our collective future.

The Vision brought by the Dream Horse inspires me to this day. It presents a possibility of what our future can be like. I know too that it will be hard-won. There will be more loss, more suffering, till that critical mass is reached. Then finally, we will say "*enough.*" It is time now for "younger brother" to grow up and to do so in haste. There is not much time left.

Ironically, as fate will have it, our actions of damaging the Earth, and the protective layer that once cocooned our metamorphosis, will lead to an uncomfortable quickening. Our situation is accelerated by invisible cosmic forces we are just beginning to understand. We have opened up a hole in the atmosphere, allowing the cosmic mother to touch us through and through.

The realm of science and material mastery is colliding with the invisible realm of spiritual insight and cosmic truth. We must learn how to ride this new wave of becoming by embracing both worlds equally. We must not fear, nor second guess, the invisible. We must come to recognize it as our own. We must let go of all we have known, or thought we knew, and reach for the mane of the flying horse to bring us safely to land. To land in a new world where people listen to the song of a skylark and consider it "better than all treasures."

25. Fourth-Generation Monarch

In our every deliberation, we must consider the impact of our decisions on the next seven generations.[1]

IROQUOIS MAXIM (CIRCA 1700-1800)

The plane glided low over Padre Island, a narrow river of sand between the gulf's blue waters. The sun rays were warming the yoke, and my fingers gently holding it to guide our direction. Life arises where the sea meets the land. This is true for Padre island, now home to the extremely rare Kemps Ridley sea turtle. The Kemps has been teetering on the brink of extinction since people started over-harvesting eggs at their former largest nesting site in Rancho Nuevo Mexico. In a compelling act of cross-border partnership, Mexican and United States conservation agencies agreed to protect the turtle's eggs and hatchlings in Mexico and reestablish a new colony on Padre Island.

From 1978-1988, eggs from some of the remaining Kemp's nests at Rancho Nuevo were collected, packed in sand from Padre Island, and transported to a special facility where the eggs were incubated and kept safe. Once the eggs hatched, the hatchlings were released on the beach at Padre Island and allowed to crawl into the surf. They were then recaptured after a brief swim and taken back to a special facility where they were fed and cared for until they were about one year old and at least eight inches long. Once turtles reach eight inches, they have a better chance of survival. The public is then invited to watch a celebratory re-release of the little turtles into the ocean surf. The entire process proved to be successful in imprinting the hatchlings to Padre Island's sand, so that the females would one day return to lay their eggs safely. After patiently waiting for eight years, the conservationists were elated when the first batch of now adult female turtles returned to Padre Island.

New turtle mommas that were released as 8 inch long babies now lay thousands of eggs to hatch the next generations of tiny turtles. So far, hundreds of new nests are found each year on Padre Island, and the eggs are still carefully removed and incubated until it's time for them to hatch. Then, several times a year around dawn, the hatchlings are released and allowed to scramble down into the white foaming surf to take their chances in the wide ocean. These release dates are one of the most treasured events at the seashore near Corpus Christi. Families come out to watch as the tiny hatchlings enter the water, to cheer for each one.

While the Kemps population is still low, and the species is still considered endangered due to other impacts on its population (such as global warming), the international cooperation and conservation efforts are an excellent example of human ingenuity in respectful partnership with Nature. As the plane circled wide over the narrow tip of Padre Island, I smiled as I considered the wonderful project and how people and agencies from different countries came together to save a species. I know our children and grandchildren will join with others worldwide to do similar good and meaningful works. These cooperative, eco-friendly projects will be on the rise with the new millennials.

~

I believe this to be true based upon my reading of the time that we now live in. Humanity is influenced by cosmic forces, the stars and even the elements. The ancients tell us that the day and time people incarnate on the planet can foretell their characteristics, personalities, and even motivations. Astrology is one method for making such determinations based on how the planets influence us. The planets, to the ancients, were considered deities that influenced earthly life and played a role in cosmological happenings.

I use a system used by ancient peoples from Europe, including Austria, Belgium, France, Northern Italy, and the British Isles. This includes my ancient ancestors, the Celts/Picts, who lived and thrived in many countries. For example, the ancient Celtic site at Hallstatt, Austria, discovered in 1846, gives a snapshot of the cultural breadth and depth of these ancient people who resided in this area for at least 4,500 years if not more. Hallstatt is considered an important economic and cultural hub for the Celts, where they managed one of the world's largest salt mines at the time and were excellent artisans and builders. The complexity of their buildings, beauty, and artistry of everything from jewelry to cooking ware and horse trappings, and obvious signs of their advanced spiritual ceremonies, place their civilization at least as high as Rome and Greece. They developed many advanced technical, scientific, mathematical, and spiritual insights and skills.[2] These ancient people had a much more holistic experience of these fields, which we deem to be totally separate in our age. One of their beliefs pertained to the influence of the five elements at the time of a person's birth.

They blended mathematical principles and spiritual insights to determine that the day, month, and year of a person's birthday corresponded with specific elements. My grandmother hinted at some of this when we sat in her parlor in Edinburgh having afternoon tea. Yet, the system is complicated and involves a great deal of information now lost to the ages as the Celts did not tend to write down their knowledge but depended upon oral transfer across the generations. In days gone by, people probably studied this system for many years before becoming competent interpreters. While no one on the planet at this time may know the system in its entirety, some of this ancient knowledge is beginning to come to the surface.[3] For example, I created an Ancient Celtic Trilogy reading that interprets three influences upon a persons character and motivations. The Trilogy reading looks at the impact of the five elements; Earth, Air, Water, Fire and Aether, at the time one was born, and combines this with an understanding of one's alignment with a spirit animal, and one's handwriting.

The book *The Code: Unlocking the Ancient Power of Your Birthday* by Johanna Paungger and Thomas Pope is groundbreaking in its depth of knowledge about this now secret ancient system. It's a tremendous resource for unlocking the meaning of your birth date by numbers. While many aspects of what Paungger lays out in *The Code* partially reflects the system I heard about many years ago from my grandmother, there are a few minor differences. For example, I was told that a person's date of birth was also associated with animals admired by the ancients. In my Celtic trilogy readings, I include information based on the animals aligned with a person's birth moon.

Finally, the third "arm" of the Celtic Trilogy Reading is an analysis of one's handwriting. This is needed because while we may be born with specific traits and characteristics, our life experiences distinctly affect us and our outlook. For example, parents may tell a child born with natural musical talents not to study or practice music because it is not a financially secure career path. Or perhaps a child with musical talents is overshadowed by an older sibling. Or perhaps they are born to wealthy parents who provide the best instruments and instructors for that child to excel.

By accurately interpreting one's handwriting to determine individual perceptions about self and the world, one can discern whether they are using the gifts they were given at birth or are hiding the gifts they were born with. This three-prong approach makes for a very in-depth analysis.

It is possible to apply, in very general terms, characteristics about entire generations of people born into different millenniums. For example, the nineteenth century, or 1800s, was largely associated with Romanticism. Romanticism referred to emphasizing emotions, intuition, and valuing the beauty and harmony of nature. It was also a time of embracing liberalism, rights for individuals, and social welfare. This aligns not only with the visionary and forward-looking qualities that come with the dominant numbers 1 and 8 of that Century and the element of air, but the compassionate and intuitive aspects of the element of water. Both of these elements, water and air, exert an influence on those born in that time period.

However, while each era presents energetic influences, many individuals can and will have extremely different propensities than the era they were born into. These individuals think and perceive things differently from the "crowd." In this way, they are "set apart" and may be said to be "ahead of their time," "old fashioned" or, in my case and the case of many people born in this millennium with its emphasis on materialism as "too sensitive." With the foremost understanding that generalizations about personal characteristics based on one's era is limited by other driving forces, we can still glean some underlying trends. For example, people born in the twentieth century are influenced by different forces than those now being born in the twenty-first century.

The generation of people born in the 1900s tend to be visionary, with great powers of persuasion and tenacity to overcome many obstacles in their path. It is fitting that many technological advancements have burst upon our world in astonishing swiftness have come about in the last 100 years under the driving influence of those born in the 1900s. Twentieth-century generations have a thirst for growth, advancement, and setting and achieving goals. They have a drive to "succeed." The impetus for constant growth, the development of the material world, and attention on an accumulation of material things are drivers for those born in this era.

The 1900s generation were born in an age that exemplifies mastery over the material realm of reality and are influenced to mold the world as they see fit. This is the age that encourages the builders in society. Yet, the epoch can support a tendency to misinterpret the cosmos as primarily a place of physical reality and ignore its spiritual and energetic aspects. The 1900 generations may apply "band-aid" cures and remedies that treat the physical symptoms but miss the underlying cause of illness. A lack of understanding about the need for emotional and spiritual balance may lead a person born in the 1900s generation to ignore others' needs, misuse the environment. This can lead to inequities in society and the overuse of natural resources to promote economic growth and rampant materialism. While people are no longer born in the 1900s, the vast majority of those in leadership positions today were. This means that the influence of the 1900s generations is still significant. While many good things have come from this era, an overriding drive to achieve material success has led to massive environmental destruction.

Fortunately, many society members were born in the 1900s who stand out from overriding generational trends. These individuals can feel like they do not belong on the planet; however, they are sorely needed. They are major contributors to civil rights, humanitarian causes, alternative health solutions, environmentalism, free energy, and spirituality, to name a few of the areas in which these individuals lead and contribute. They are way showers, to help guide humanity on its path towards self-actualization.

The twenty-first century (2000) carries very different qualities than the twentieth century (the 1900s), especially since it's a brand new millennium. The qualities of the 2000 epoch influence those born during this time to consider their actions on others. "Inclusiveness" is an important driver and leads those born in the new millennium to work closely with people of different nationalities and races. Teamwork and community are overriding goals. Those influenced by this age can also be more attuned to the environment's needs and the needs of other life forms. This era is also a time of taking action and making change happen quickly. It epitomizes "striking while the iron is hot" and "seizing the day." Those that exemplify this time will inspire with their actions and bold oratory skills. This is why those born under the influence of the 2000s are already influencing our world, even though they are still quite young.

The era of the 2000s generation is now upon us. This new generation needs to be careful not to fall into the habit of taking on too many projects or not seeing them to fruition. There is also a danger of investing their energies in shallow pursuits when more substantial action is needed. They may wish to escape into fantasy, or the internet, versus tackling the changes the world needs. Those of the twenty-first century can benefit from their elders who hold historical knowledge. Yet, given the state of the world, bold actions are indeed needed now more than ever before.

The generations of the New Millennium will make sweeping changes, some of which have been initiated by their forward-thinking elders. Yet, the world will not fully feel the impact of this new generation until they command leadership positions in society. We are already experiencing the fiery new generation's boldness and inclusivity as young as 15 and under have stepped into roles previously held by those decades older.

For example, Greta Thunberg, a 16-year-old Swedish activist, is a perfect example of a 2000s generation leader. She has a burning drive to stop global warming and went alone to protest outside the Swedish parliament during climate talks when she could not get anyone else to go with her. There she announced her "School Strike for the Climate," a movement to ask students to walk out of schools and classrooms to demand a more sustainable future. Thousands of students in nearly 300 towns and cities worldwide have joined her #FridaysForFuture protest, where students walk out of their classes one day a week every Friday.[4] Today, Greta has lit a fire under people young and old to take action.

Similarly, the Parkland High School students Jaclyn Corin, Alex Wind, Emma Gonzalez, Cameron Kasky, and David Hogg, working to stop gun violence, led a "March for Our Lives" on March 24, 2018. It is estimated that over 2 million people joined the March from different locations around the country. It turned out to be one of the biggest youth-led demonstrations since the Vietnam War.[5] Inspiring speakers, they have also attracted many to their cause.

In *Juliana v. the United States*, 21 young people from the ages of 11 and up sued the United States, alleging that their constitutional rights have been violated by the government's policy of burning fossil fuels and destabilizing the climate.[6] They argue that the government's failure to address global warming is a violation of the public trust doctrine that requires the government to maintain a livable environment and the youth's fifth and ninth amendment rights to "life, liberty, and property." They seek to have the United States government prepare and implement a "science-based climate recovery plan" to return the CO2 to 350 parts per million by 2100.

~

A momentous change that will take place in this new millennium is the acknowledgment that the human race is not alone in the Universe. "Earthlings" will learn (what many already know) the startling truth that we are not the wisest nor most evolved species in the Universe. (The US Department of Defense has already released three declassified videos of what they refer to as "unexplained aerial phenomena."[7]) This revelation, perhaps the greatest to rock human society, will sweep away everything we thought we knew about our place in the Universe and our position in the web of life. We will then see our specie's present-day naivety and misunderstandings in much the same way that we now look back with disbelief upon those days when we alleged that the sun revolved around the Earth.

We will no longer be able to consider *Homo sapiens* the pinnacle of intelligent life in the universe. From this place of greater equanimity and humbleness, humanity will finally be able to appreciate and value different life forms on our own planet. This is because we will measure ourselves in relation to other Earth species on a more equal footing, with neither them nor us being the highest nor the lowliest of creatures. We will also comprehend that our Earth-island floating in the vast sea of the Milky Way is our very own living Gaia that shaped us and "mothers" us. While it's clear this shift is already unfolding, it's coming at a very fortuitous time—the twenty-first century with its emphasis on others, community, and relationships. It is time for the Earthly human race to join our cousins in the stars.

The millennial cycles that affect human consciousness and evolution are held within much more significant cycles known as the Great Year. The Great Year is approximately 24,000 Earth-years long. This is also known as the precession of the equinox. This is caused by the movement of the sun and our entire solar system spiraling through the galaxy.[8] Over thirty ancient cultures refer in their own way to this Great Cycle. In fact, this cycle was better known in ancient days than today, as reflected in the stories held by ancestors from many cultures as diverse as the Polynesians, Egyptians, Sumerians, Mayans, etc. The Romans referred to these cycles as the Golden, Silver, Bronze, and Iron Ages.[9] The ancients believe that there was a golden time on Earth, which some refer to as Atlantis's time, when humankind was at a zenith. The Hindus refer to the cycles of the Great Year as the Yugas. They believe that there are four great Yugas:[10]

Kali (the first and lowest age)
Dwarpa (the second age)
Tetra (the third age)
Satya (the truth, or the final golden age)

Sri Yuketswar, an Indian monk and yogi, and a noted astronomer determined that we are now in the Dwapar Yuga, the second age, and have been for over 1,500 years after leaving Kali's lowest age. According to Sri Yuketswar, we left the lowest age and entered Dwarpa Yuga in 1699, and this is reflected in the great and rapid advancements in science and technology. However, since we are still at a somewhat lower stage, or "Yuga," there are many things that we do not understand. Thus, our world's present unhappy state, our use of harmful technology, and modern man's focus on material "pleasures" versus spiritual gains.

Vedic beliefs state that, along with our Sun, which brings us life and evolves our consciousness, there is a much greater "central sun" in the middle of our galaxy that bathes the solar system and Earth in "celestial light," nourishing and propelling forward the evolution of consciousness of all beings within.[11] When our planet is closest to this central sun, human consciousness will be at its highest capability. People's ability to know life, communicate and be understood, and live happily and harmoniously with others on Earth will be at its zenith.

This ancient belief is mirrored in the scientific discoveries of Paul LaViolette, who developed the galactic super-wave theory and argued for the existence of a central galactic sun. The Starburst Foundation illustrates LaViolette's concept of the central sun thus:

> *The luminous cosmic ray emitting source at the center of our Galaxy is a celestial orb about 4.3 million times the mass of our Sun and the most massive object in our Galaxy. Currently, it is seen to radiate about 20 million times as much electromagnetic energy as our Sun. By one estimate, it radiates most of its energy as cosmic ray electrons and protons, giving it a total luminosity 2.5 billion times that of the Sun (LaViolette, Sub-quantum Kinetics, 2012). Based on early radio observations, it was given the designation Sagittarius A.*[12]*

The cycles within the Great Year or Yugas are said to have two distinct and equal halves. Whenever our solar system comes closer to the "mother sun," or Sagittarius A*, all the creatures living in our system rise to greater possibilities. Whenever our solar system moves away from it, life falls to the lowest level of possibility. When our solar system is closest to the "mother sun," Satya Yuga will begin; when it departs and moves away from the "mother sun," we enter the lowest age, Kali Yuga when consciousness and understanding are at their lowest levels. This could then explain why modern man is now catching up to "ancient" advanced knowledge.

We will enter the next age of Tetra in 4099 C.E. and remain there until 7699 C.E., at which time we will enter the highest, or golden, age of Satya Yuga. At this time, all people will be so advanced that we will be able to communicate telepathically, travel to space, and conduct many other activities we would deem impossible today. People like Nicolas Tesla have tried their best to move our knowledge about cosmic energy and intelligent life on other planets forward, despite resistance from those tied and addicted to the material world. There are already a growing number of people on the planet who are tapping into greater cosmic consciousness and are "way ahead of their time."

Contemplating the great cycles, I have come to a clearer understanding of the cosmic Vision brought by the Dream Horse. The cosmic ribbon of light that touched the Earth's surface was sent from this central star, the mother sun of the Milky Way Galaxy. This ribbon, containing celestial light from the dawn of existence[13], is bathing the Earth to "soften" and evolve the consciousness of humankind. If one considers the Earth to be a living being, just imagine the great wisdom and vast consciousness of the central sun, *2.5 billion times the sun's size in our solar system.*

The area known as Sagittarius A, in the center of our galaxy has been misinterpreted by some to be a black hole, that devours mass and light. However, the black hole theory is being and has been disproved. Light does escape from Sagittarius A. In fact, a great deal of it!

Three and a half million years ago, Sagittarius A unleashed an enormous burst of energy. The light from this gigantic flash illuminated the Magellanic Clouds outside of our Universe. This cataclysmic event affecting everything within and locally near the Galaxy would have caused a glow to emanate in the night sky above the Earth for up to a million years. Our ancestors would have seen this light shining appearing suddenly, brightening the very center of the Milky Way Galaxy and beyond.

Around the same time frame, the fifteen different hominid species that we are aware of, and possibly more, narrowed to just five. As the light continued to shine forth from the galactic center, homo erectus, our common ancestor, arose three million years ago, in the wake of this cosmic event.

Recently, Sagittarius A released another burst of light that increased its brightness by 75% in May 2019. This burst lasted over 2.5 hours. These cosmic emanations are very powerful and intense. Sagittarius A is a womb in the center of our galaxy creating light and even mass through a vortex motion. This is possible because all mass is born from the "invisible" Aether.

Much of what western science has told us about our Universe and universal laws is wrong. In this materialistic age, scientists have seen only what interests them, and largely ignored or misinterpreted the power of the invisible. By doing so they have avoided and ignored critical aspects of how the cosmos works. Namely mass arises from the seeming invisible. The electron, the smallest whole particle of existence, exists in a mass-less state until the Aether moves upon it causing a vortex from which mass can be created. The primary source of cosmic energy arises from the dynamics of living Aether and how it "dances" with the smallest particles of creation to produce fields, such as gravity and electricity, and ultimately even matter itself.

The spiral flight of the osprey holding the wide eyed fish opened my eyes to this universal truth; motion acting upon the invisible, is the creator of our world, and many others.

Thus, the flash of the cosmic ribbon of light that I witnessed was both a physical occurrence and a spiritual event emanating from the mother sun. This flash will rock human societies and cause suffering to the human race and especially those clinging to base materialism. Yet, it is needed more than ever at this time as human societies endanger our world, humanity, and the lives of all beings. The cosmic event ushers in a time of great spiritual insight and upliftment for those who are ready.

~

One of the most visually stunning sights in the world occurs in the Michoacán hills in Central Mexico. From October to March of each year, millions (sometimes billions) of gloriously vibrant orange and black butterflies, *Danaus plexippus*, otherwise known as monarchs, fly to a special forest there to hibernate on their favorite trees. After passing through small towns, villages, and wild places in Mexico, the butterflies, weighing in at only .50 grams on average,[14] climb to a high altitude, eventually landing on the sacred oyamel fir tree nearly two miles above sea level.

The oyamel fir grows in the brisk air between 9,500 to 11,000 feet in elevation.[15] It has an upwardly pointing crown that some say resembles praying hands. It is on these iconic trees that the monarchs make their winter home. They cling in large groups by the thousands, covering the tree-like a bright orange and black living garment. Many who have visited these trees when the monarchs are in residence are left speechless at the beautiful spectacle. Those who have visited the forests say that the subtle movements of the monarch's wings sound like a gentle rain.

These butterflies, which start their long journey in North America, seek to arrive in this region and reach a specific individual tree. They prefer to over-winter on the very same tree that their great grandparents chose. They follow the wingtips of their great grandparents, not their parents, grandparents, or even great grandparents, because only every fourth generation can make the astonishing 2,500-mile journey.

Not only is this fourth-generation monarch endowed with an expansive life span, compared to the three generations before it, but it is graced with navigational abilities and greater wing strength. Flying at speeds up to 25 miles an hour and riding thermals to help them along the way, these joyful, colored beings fly from the United States or Canada through Mexico's Trans-Mexican Volcanic Belt region.[16]

Their arrival in October in Mexico is celebrated by local communities who honor them as their human ancestors' returning spirits.[17] Many cultures, like that of the Aztecs, have had similar beliefs about butterflies, referring to them as incorporating human souls.[18] Even Aristotle referred to butterflies as *psyche*, the Greek word for soul.

Meanwhile, Mexico's residents never see generations one through three since they rarely live past their sixth week after emerging from the pupae stage thousands of miles away. Six weeks gives these butterflies just barely enough time to feed, mate and lay eggs, and perhaps dream of the fantastical voyage of their descendants in the future. For it is their descendants, and only the fourth generation will far exceed anything that they could be or do.

When it comes time to produce the special generation four eggs, generation-three parents must infuse them with great longevity, strength, and passion for the continuation of the species. They are the proud parents of what will be a super-generation, a generation that can live ten times longer and fly thousands of miles further than their parents. Compared to a human life span, it's like parents giving birth to a "super" child capable of living to the grand old age of 700 years!

Sadly, today, monarch butterflies are declining, primarily from habitat loss, pesticide use, and especially the milkweed loss, considered an undesirable weed plant by farmers who apply a fungicide to remove it.[19] Some organizations engage in educational campaigns asking people to plant milkweeds and/or not destroy those growing to provide habitat for the monarch eggs and caterpillars. Yet, that is only part of the solution. Due to logging and extreme climate changes, which are making the world hotter and drier, the sacred oyamel firs are disappearing.

The loss is happening quickly, with a possibility that up to 70 percent of the monarchs' sacred forest habitat will be destroyed in the next ten to twenty years.[20] While trees can migrate to more beneficial areas over many generations and have done so in Earth's past, they are completely helpless at the lightning pace of climate changes today. Some scientists and ecologists are scrambling to plant seeds and seedlings of these special fir trees at higher elevations to expand their range into the cooler, wetter territory. According to Cuauhtemoc Saenz-Romero, a forest geneticist at the Universidad Michoacana, "*We have to act now, later will be too late, because the trees will be dead or too weak to produce seeds in enough quantity for large reforestation programs.*[21]"

The monarch's story is a compelling and aligned story for the potential destiny of the human race. The human race, along with the Monarchs, are now in a race against time to survive. Will it be Greta Thunberg's generation that breaks free of societal limitations and learns to live in harmony with the natural world, or will it be the next? One thing is certain: we cannot wait. We must act now to save not just oyamel fir trees and monarchs but to protect the future for humankind and the Earth.

We live in an unending and unlimited sea of energy that surrounds the Earth. One of our first actions must be to utilize innovative ways to obtain free energy from the Aether and power the world with life generating forces. Our descendants will certainly be tapping directly into this "living sea" to power their world with unlimited, free and non-polluting energy. Machines that do just this, have already been invented and independently tested. These new inventions just need the political and economic support to bring them to humanity. It is these "gentler" ways of producing energy that will help to free humanity and the Earth from overriding outdated and destructive technologies that are over a hundred years old. Once humankind resolves the problem of free, safe and sustainable energy for all, we will be able to focus our attention on discovering deeper spiritual truths, creating a more beautiful world, and strengthening our relationships with others.

Our descendants may not live to be 700 years old, but hopefully, they will be thriving 700 years into the future. If so, they will be thoughtfully considering the impact of their actions on the Earth and future generations. They will understand that their very thoughts contain great power. Like the Kogi, they will understand that humans have a role to play in envisioning our world. A world where the material realms and the spiritual realms meet. These descendants will be like Monarch Generation Four, taking the human race forward to a beautiful and unimaginable future.

26. Upon Landing

Now he who plants and he who waters are one, but each will receive his own reward according to his own labor.[1]

CORINTHIANS 3:8

It was late morning, and the Katana's 145-horsepower engine was holding steady at 104 knots. The loud crackling voice of the air-traffic controller eclipsed the engine's roar. My destination, a small airport on the outskirts of Port Isabel, thirty miles north of the Rio Grande River, was coming into view. As the plane circled clockwise and began to lose altitude, the view of the airfield filled the windshield. The narrow runway had been newly resurfaced with asphalt. It would be an easy landing. As the plane lowered gently, the light crosswind buffeted its sides, causing the plane to shift rhythmically and slightly from side to side.

As my sky-high view narrowed, my musings came down to Earth with the plane's descent. All the cares and concerns that had been kept at bay while in flight began to crowd my thoughts. I worried about the state of the Earth, the animals, and plants; how we are leaving the world to future generations. As the wheels touched down on the airfield runway, I made out a familiar figure near the small airport building. It was Jimmy Paz, the Sabal Palm Sanctuary Manager, waiting to meet me. He was wearing his brown leather jacket and standing with hands in his jean pockets. His dark curly hair blew this way and that in the wind.

Jimmy is one of Audubon's best sanctuary managers, in that he cares equally for the people and the environment. Jimmy succeeds in meeting the sometimes-uneasy balancing act of engaging people while protecting wildlife. For example, former sanctuary managers wanted to keep people out of the sanctuary to keep it as pristine and undisturbed as possible.

One year, a local community member posed this hypothetical question: "*If there was one last endangered plant left, and we wanted to use it for keeping our cultural traditions alive by using it to make a sacred object, would you let us have it?*" In other words, what was more important, the continuation of human culture or the survival of the rare plant? Fortunately, we never were faced with this dire situation. Yet there were portions of the sanctuary that Jimmy managed where no one was allowed access except for staff. This included Green Island in its entirety--one of the "jewels" of the coastal sanctuary system.

However, unlike past sanctuary managers, Jimmy always thought about his neighbors' needs as well as caring for the hundreds of species in his care. This is why he created community-based wildlife education programs for adults and youth and hosted community events. The sanctuary became a welcoming destination for the locals. Adults and children enjoyed the beauty of the place, saw animals they may never have seen before, and learned about species in their own backyard.

While there are places on the Earth that must be left undisturbed for wildlife, and part of the sanctuary was also off-limits, there is great value in engaging people to discover and be in Nature. Jimmy largely got it right in understanding that we must allow people to experience Nature firsthand and love it before we can expect them to know how to care for it, or even want to.

I readied the plane for landing, slowing my speed as I approached the runway. Once landed, I taxied over to the small airport building. As the plane came to a stop in front of the terminal, Jimmy smiled and waved. I waved back, took the key out of the ignition, opened the cockpit door, and stepped out on the tarmac. I stretched and felt the heat of the sun warm my back as I gathered my briefcase and backpack of clothes.

As I walked towards Jimmy, I anticipated his response to my smiling greeting of "*Good to see you, and how are you?*" Jimmy is not the type to complain about his job's challenges, which are many, or that just yesterday, he was up to his waist in muddy water, fixing the pump to keep water in the Resaca. Instead, he will say cheerily, "*doing great, and getting better all the time.*" Jimmy is an optimist. It's optimism grounded in practicality and love.

Jimmy is applying himself to what I call "Good Works." Good works refers to a field or occupation beneficial to the environment, or that provides a valuable service to humanity. Thus, Jimmy's optimism is very practical. Things are getting better because he is applying himself to making things better. Most people doing "Good Works" are dedicated to their jobs as an act of love. It does not mean that they do not have to deal with daily challenges, "doing good" outweighs frustrations they face.

A few other innovative programs that fall in the category of providing "Good Works" are the Non-Human Rights Project, Tree Sisters and Archangel Ancient Tree Archive. The Nonhuman Rights Project is the only civil rights organization in the United States dedicated solely to securing rights for specific animals that have been designated as "conscious beings." Animals like apes, chimpanzees, and elephants for example, are the focus of their campaigns. They work through litigation, legislation, advocacy, and educational efforts to change the common law status of certain species to "legal persons." They assert that intelligent animals have a right to specific fundamental rights such as bodily liberty and integrity.

They have helped to shine a light on how keeping intelligent animals like these alone in cages most of their lives, for example, is akin to solitary confinement for a human being. Presently, animals have no rights and are simply treated as possessions and objects. This then allows their "owners" to do with them what they will. If it's a kind and generous "owner" the animal will be treated well but if not, there is little protection for the animal. The NonHuman Rights Projects' approach of bringing groundbreaking lawsuits, with the support of world-renowned scientists like Jane Goodall, seeks to establish rights for animals and uplift them from being deemed mere possessions. Their work moves humankind forward in our understanding that animals are conscious and sentient. We must treat animals as self-aware, autonomous beings not just out of concern for their welfare, but out of respect for legally enforceable rights. This way, we do not have to depend on the intentions or actions of each "owner" to "do the right thing" in the hope that they treat their animals well. Instead, it's a legal requirement.

Then there is David Milarch, who founded the Archangel Ancient Tree Archive, and worked in his family's nursery business for over two decades. He had a wonderful wife and two sons he loved, but he was dying of renal failure because he was an alcoholic. Rushed to a hospital emergency room with fluid in his lungs, the doctor on staff told him he only had a day or two to live. A day after David was sent home, he did indeed die. As his spirit floated up to the ceiling, he looked down at the bloated, abused body below and felt he had wasted his life. David then saw a beautiful female angel in a white gown emerge.

The angel took David into the tunnel of light. At the other end of the tunnel angels told David he had work to do on Earth, and he was sent back into his body to live and complete his mission there. Some months later, after David had started to recover and get out of bed, he felt the angels near again, this time with a clear plan that came through to David by way of automatic writing. According to David,

> *My heart raced as I read through what I'd written. The earth's trees and forests are getting sicker, weakened by pollution, drought, disease, and bugs. I was to clone the biggest, strongest, hardiest trees, trees that had lived hundreds, even thousands of years, so the giants of the forest could one day restore the world to its natural order. I felt like Noah; a simple man told to become a shipbuilder and a zookeeper.*

Within a year after this divinely inspired mission, David launched the Archangel Ancient Tree Archive. Its mission is to propagate the world's most important old-growth trees before they are gone, reforest the Earth with the offspring of these trees, and archive the genetics of ancient trees in living libraries worldwide future.

I spoke to David on the phone when he and his team came to California to take fresh genetic material from giant sequoias in the Sierra Nevada mountain range. These trees can grow up to 250 feet tall and have trunks as wide as 30 feet in diameter. Once the fresh growth suitable for cloning is captured, it is carefully cultivated in a special fluid and then planted in the soil when its roots take hold. Once it is large enough to have a fighting chance of survival in the wild, David and his team of scientists select an optimum location for planting the seedlings. David said that "*With only 10 percent of the world's oldest trees still surviving, this is one way to ensure that the genetics of the longest-living trees is carried forward.*"

Meanwhile, across the waters in England, TreeSisters is empowering women at a deep spiritual and emotional level with Nature's restoration. TreeSisters, according to its founder Clare Dubois, was "*birthed in light of humanity's current circumstance.*" And from an act of love. Clare was driving too fast and did not see that she was coming up to a patch of clear ice. She lost control of the car, which careened around, bounced off the side of the road, and then headed straight for a cliff. Fortunately, a lone tree blocked the car from descending over the steep edge to her probable death. As Clare hit the tree, she saw a blinding flash of white light, and two words lit up in the light.

Then Clare heard a voice: '*You have to reforest the tropics within 10 years!*' This seemed like an impossible mission to Clare. She was being asked to take on a huge task and be bold; this would require a shift from her being a loner to being a leader, from being an introvert to being an extrovert. She was being asked to become someone unimaginable to herself. Yet, the voice persisted: "*You have to mobilize the women; the women are the missing piece, the women are womb carriers, and Earth is a womb. Women and nature carry the same history; what has been done to a woman has been done to nature . . . that feminine consciousness was the wisdom of living systems, of seasons, of giving and taking not just taking . . . that humankind had normalized taking from the planet without thought for consequences and now needed to normalize the act of giving back. Women have to remember who and what they really are. Feminine consciousness needs to be reinstated.*"

Today TreeSisters is in over 50 countries and supports the planting of almost 1.8 million trees a year across eight ecosystems in six countries. This includes mangroves and dry deciduous forests in Madagascar, cloud forest, and agroforestry in North and South India. They are even reforesting the southern and northern slopes of Mount Kenya to replenish that crucial water tower and reforesting around the last remaining 250 critically endangered Cross River gorillas in Cameroon, and growing forest corridors in the Brazilian Atlantic Rainforest. When we last met, Clare was still mourning the loss of trees she witnessed in the United States. While Clare feels she does not have all the answers, but she believes that "*We do not know where we go from here, but as a TreeSister, you align with humanity's shift from a consumer species to a restorer species and a future where we give back to nature.*"

~

As Jimmy and I leave the airport to drive to the sanctuary, Jimmy speaks proudly about how well the local youth are doing in his new community program. Jimmy educates them about wildlife and ecology as they work to restore the sanctuary; planting trees, putting up bird blinds, and opening muddy waterways to restore a healthy flow. He then switches topics to his recent sightings of rare and unusual animals, like the scissor-tailed flycatcher, or the chachalaca, with its staccato song rising from deep in the tropical Sabal palm grove. Or how he even got a glimpse of its brown feathers behind the fanlike Sabal Palm leaves.

As he drives, Jimmy gives an update on how well "his" birds are doing on Green Island. He took a trip to the island just a few days ago to make sure that the colonies are safe and thriving. He knows how many nesting pairs are settling in and how many eggs and babies are in their nests because he counted them all. As I looked forward out the windshield towards the long narrow road leading to the sanctuary, the anxiousness I was feeling upon landing lifted and I smiled. There is no better way to feel than knowing that new life is coming into the world under the watchful eye of someone, like Jimmy, who cares.

Ultimately, we must not only apply ourselves immediately to save our world through traditional means and actions. We must learn to "see" our world with new eyes, and gain knowledge about the "invisible" forces surrounding us every day. We will be continually astounded by the vastness of Nature's intelligence. It is through gaining insight about the true nature of Nature that we will not only be able to save the Earth, and our own species, but we will learn that even we have abilities beyond our imagination.

Humankind can learn to understand and partner with the "invisible" verities of our world, such as the intricate forces of Nature that seek to keep all things in balance. For without this wisdom, we will not be able to stop the ecological devastation sufficiently in time to protect all life on Earth. There are millions of amazing things about our world we will one day understand. Consider the flight of a bumblebee.

The bumblebee appears to violate the known mechanics of flight. Bees notoriously lack aerodynamic qualities with their large rounded bodies and slight wings. The wings achieve some vestiges of lift by moving back and forth and creating little vortices of low air pressure. Yet, this effort does not appear to be enough to lift and sustain their large bodies in flight. However, there is an alternative explanation that highlights the invisible frequencies within the bee and around it.

The supposedly aerodynamically " bereft" bumblebees not only fly but can reach speeds up to 30 miles per hour, even fully ladened with pollen! How do they do it? They "levitate."[2] According to Arthur C. Aho, the author of "*Tomorrow's Energy: Need Not Be Fuel*," there is a hollow tube in the bumblebee's throat, and when the bee rhythmically beats its wings, this causes frequencies to bounce back and forth inside this cavity until a specific resonance is reached. Once the frequency of the inner cavity syncs with Earth's vibration, the bee becomes free from the tug of gravity. The bee and its somewhat rotund body become weightless, and its wings are then used to set direction and altitude.[3] This anti-gravity effect was also discovered separately by Viktor Grebennikov, a researcher and author. He found that some beetles shell-like wing coverings could create levitation by their precise geometric structures.[4] Equally fascinating, or disconcerting, depending on your world view, is the incredible hurricane sensing ability of Veeries.

A Veery is a small delicate brown thrush that nests in forests in the United States and Canada. Despite its delicate physical appearance, it migrates thousands of miles from Brazil to its nesting grounds and back. While the early Spring flight to Canada and the United States can be arduous but uneventful, its journey back to Brazil can be fraught with danger. Between August and October, 84% of all hurricanes and 93% of all major hurricanes hit the Atlantic basin during peak songbird migration.

Over time, scientists noticed that the birds cut their breeding season short and left early on some years, and in other years they stayed to start another brood. Most thought that this was just an unremarkable occurrence, that is, until Christopher Heckscher, an ornithologist studying Veeries, noticed that the birds seemed to be accurately predicting the intensity of hurricane season well in advance and from thousands of miles away. The Veeries predictions were uncannily accurate, even beating out "state of the art" meteorology computers. What Heckscher learned over twenty years is that Veeries not only have the ability to predict the severity of hurricane seasons, but they correctly foretell the upcoming season months in advance.[5]

How these delicate little beings manage such an extraordinary feat is a mystery. For certain, there is a great deal that we do not understand about our world. Thus, while scientific reports, charts, and predictions present damning evidence about the state of our world, and the future indeed looks grim, our rapidly evolving wisdom and insight about our wonderful, intelligent and sacred planet may offer some respite.

~

So here we have come full circle in our journey. Just as the pelican, inspired people hundreds of years ago, then lapsed into almost certain extinction, to being now protected and cared for—we ourselves have come full circle through this story. In the midst of our present and dire challenges, we need not wait for a golden age, or a time when the entire human race finally lives in peace and harmony with the planet. We can choose today to take on and do "Good Works" spurred by our care for the Earth and each other. We need not understand everything to start; we can learn and deepen our relationship to our planet and other species as we go. We will then find that our lives have taken on greater meaning, as we lovingly co-create the world with the One who made it.

27. Into The Wilds: What Coyote Teaches

"He prayeth well, who loveth well
Both man and bird and beast.[1]"

SAMUEL TAYLOR COLERIDGE

The wind-tossed rain fell in parading sheets onto the thirsty ground. My face, the only part of my body, not covered by the sleeping bag, felt the stinging-cold droplets. Squinting my eyes between the drops, I spied thousands of bright stars shining intermittently between fleeting, swollen clouds.

Fifteen years had passed since my flight to Port Isabel, Texas. It was 1 am, and I was wide-awake and surrounded by tall wild rye grasses tossing wildly in the wind. I lay in the center of a meadow, at the edge of miles of forested wilderness. At the meadow's border, blue and white oak trees were dancing a wild jig, while the tips of their branches touched each other excitedly with each gust of wind. The exhilaration of the land and sky in the storm was visceral as lightning spiked the sky, followed by booming thunder. It was as if I were laying between two lovers, father sky and mother earth, as they shook with passion in the act of Creation.

After the rain, the large full moon began to rise in the sky when not overshadowed by clouds. Its milky whiteness caused tufts of grasses to cast long shadows on the ground. As I stared into those moving shadows, I had a distinct feeling that I was not alone. Goosebumps appeared on the back of my arms and neck. I sat up and looked towards the deepest grove of trees. Standing at the edge of the forest was a male coyote about four and a half feet long, with a light-colored chest, finely chiseled snout, and large ears. He stood with front paws planted firmly on the earth.

Sometimes called "prairie wolves" because of their fondness for flat terrain, coyotes are native to American soil. Here they have successfully remained despite the urbanization of much of their former habitat. Originally found only in the West, coyotes have extended their territory today to as far east as Maine. Wily survivors, they also appear to have interbred with wolves to create larger, fiercer coyotes. An animal mistakenly called a "red wolf" (which DNA testing shows to be more coyote than wolf[2]). Immortalized in modern culture as the wily coyote, and Coyote the Trickster, the coyote is often the centerpiece of many indigenous peoples' stories, including the Coast Miwok, who once lived in great numbers where I live now, in Northern California.[3]

Coyotes are one of the few wild carnivorous creatures that have adapted well to human society. They often live near and, in a few cases, within cities. This was the case with Adrian, the coyote who entered a Chicago Quiznos at lunchtime in downtown Chicago.[4] Yet despite its geographic proximity to humans, the coyote remains an enigmatic and mysterious creature. Coyotes are unlike their cousin, the gray wolf (the ancestor of all modern dog species, from boxers to toy poodles), in that they cannot be bred for domesticity. By the third generation of a domestic breeding program, coyote offspring show up infertile or genetically deficient somehow.

The coyote stood within the protection of the shadowy trees. His tail and hindquarters were hidden in darkness while his chest, head, and front paws were clearly visible in the moonlight. I felt his presence as if he were right next to me. An ancient wariness filled every fiber of his being. I found it remarkable that I could clearly see his deep yellow eyes at that distance. I had the distinct feeling of being carefully sized up.

He lowered his head slightly and lifted his left paw off the ground. As if summoned, I felt it was time to move. I pushed the sleeping bag down off my body, unveiling my legs, and stepped out of its heavy, draping form. Barefoot on the cold, wet earth, the ground's touch caused tingling to sweep through my body. I felt invincible to the cold and the rain after being out in the weather for most of a week.

I saw my hiking sneakers glowing pale in the moonlight, still lying where I had carefully laid them just within arm's reach. Yesterday morning I had awoken to find one missing. After searching for a good 30 minutes, I found it had been carried off some distance towards the woods. *No doubt this was Coyote's work,* I thought to myself. Luckily, tonight both sneakers were still there. I leaned down and placed my hands on their cold wetness.

Six days ago, I had come to one of the largest wilderness areas in California to lead a Vision Quest. I had been leading Vision Quests to this location for almost a decade. During Vision Quests, participants face their fears, deepen their connection to the land, and tap into inner ways of knowing. For the first days of the quest, Vision Quest participants receive instruction in animal and plant identification, tracking, dream recall, reading the landscape, and working with energy in the land and plants. They also learn how to meditate or how to deepen their existing meditation practice. They also learn how to create a sacred circle and conduct their solo. During which time, they spend four days and nights alone in Nature.

Over the years, I had come to know the animals and the trees in the area very well and recognize many of the individual coyotes, bobcats, wild boars, foxes, turkeys, as well as individual animals of other species. I also had my favorite trees:

The blue oaks, with their silvery green leaves

The manzanitas, with their smooth, red, flesh-like bark

The tall, stately firs.

In the predawn hours of yesterday's early morning, the twelve Quest participants had packed up their gear, removed their sacred circles, and walked back to base camp. They shared their remarkable solo experiences, and after breakfast and amidst hugs and tears, left to drive home. I now lay alone in the middle of the night on a high open field above base camp. But I was not really alone; Coyote was here.

I had keenly felt Coyote's presence when the group arrived in the afternoon on Monday when the sun was warm and high in the sky. After teaching the group about the resident flora and fauna and how to sense energy in the land and plants, we settled in for evening dinner and a shamanic spirit animal journey. Later that night, just before falling fast asleep, I had heard a Coyote howl three times close-by.

Now, as I stood alone in the soft moistened Earth, Coyote stared at me expectantly and then turned and faced the forest. Before taking off, he briefly looked over his shoulder to make sure that I was following. I quickly started walking in his direction as he took off and disappeared into the woods. Coyote was moving at a leisurely gait for an animal that can reach speeds of over 40 miles an hour. I reached the forest and still saw him moving through trees ahead. I needed just to maintain a light jog to keep up with him. I stayed about 200 to 300 feet behind him, a distance that grew slightly when obstacles like fallen trees and rocks came between us and shrank when we traveled in clearings or when he slowed down to let me catch up.

Much of the time, we traveled in a straight line, for Coyote, like Crow, prefers the most direct route between any two points. Coyote would sometimes disappear behind tree trunks or shrubs. At those moments, I was guided only by a glimpse of the black tip of his tail or the tips of his ears. Sometimes, it was just a palpable sense that he was directly ahead that led me onwards.

As we traveled, darkened clouds began to form once again, and soon tiny raindrops fell cold and sharp. I welcomed the piercing rain that awakened my senses as the clouds flew swiftly in the sky, causing the landscape to flow from shadow to illumination when the moon's face peeked through.

We made our way out of the forest into another open field. Seed heads of tall grasses brushed teasingly against my calves as we ran. Soon another forest populated by coastal live oak, California bay, and buckeye trees rose before us. The moonlight flickered between leaves and branches with the gentle pitter-patter of light rain when we entered.

A dark shape floated for an instant in front of my face and turned into the ghostly outline of a large horned owl. Its broad wings beat in butterfly-like movements as its large graceful body and wings flew between the branches and disappeared noiselessly into the night. After stopping to observe the owl, I looked towards the trail to find Coyote but had had apparently continued without stopping and was nowhere to be seen. The acute sense that had previously guided me to find Coyote had disappeared with the owl. The rain clouds had gathered, and the forest was now dark and silent. I had lost my bearings. I did not know where I was or even how long I had been running.

Childhood memories of "getting lost" in the woods because of either running ahead of my family or going off on my own explorations flooded my mind. I was no longer the seven-year-old child who ran off in the rain forests of Puerto Rico after the beautiful gray kingbird on a family vacation, or the eleven-year-old going for a quick walk into the forests in Virginia, that turned into an entire day and night sojourn. People were looking for me in both these cases, scouring the woods, and asking passers-by if they had seen a lost girl.

There was no one looking for me on this night. I chastised myself for not knowing better than to run off into the woods on a whim simply. I shivered and clasped my arms around my chest. I knew that if I got a good view of the stars, I would determine North from South and plot a direction. I walked up-slope to find a clearing, hoping for a view of the sky. I had all but forgotten about Coyote.

I soon came upon an enormous Douglass fir towering above the other trees. It was probably over 300 hundred feet in height. Protected by its broad-reaching and thick canopy, the ground beneath the fir was parched. I sat down and leaned my back against its broad, comforting trunk. The faint pitter-patter of the light rain on its upper branches was rhythmic and soothing. I inhaled the sweet, crisp scent of fir needles and felt at peace.

A sound of shuffling leaves broke my reverie. The noise, which would have been barely audible on a typical city street, was incredibly obvious in the quiet of the forest. It came from behind me, on the other side of the fir tree, so I stood up and walked around the large trunk. There below one of the trees, low hanging branches, was Coyote sitting on the ground.

He seemed to be enjoying the dry ground as much as I was. As I came around the tree, he lifted his body slowly, stretched his forelegs in front of him, and yawned. Kicking his hind legs back and forth energetically, which caused a spray of dirt to fly up, he issued several low yelps. Finer particles suspended in the air and floated in a cloud around his head that would never return to the ground.

As he performed these brisk movements, Coyote seemed oddly unfamiliar. Perhaps I was unfamiliar to him as well since he lifted his nose to smell the air in my direction as if making sure that it was the real me. He then turned away and walked to the edges of the tree's boughs. With his tail held parallel to the ground, he walked entirely around the tree as if he was looking for something. He did this two more times and then sat down where he had started and looked at me expectantly. His circular slow dance enthralled me, and when he became still, the familiarity that I had previously felt towards him returned.

Stepping ahead of a fallen branch that was partially blocking my view of him, I caused a twig to crack loudly beneath my foot. The ensuing crack seemed to break the spell of our communion. Coyote leaned his taut body forward, heightened his shoulders, which created a sharp ridge where his shoulder blades extended from his body. He looked towards me for an instant, his eyes like glowing embers, and then quickly disappeared into the woods.

I knew there was no point in following him now, as he was moving far too quickly and did not seem to want me to follow. Disappointed and tired, I decided to walk back to base camp since the rain had stopped and I could now see the stars and plot direction. I had only walked a few steps from under the fir tree when heavy drops of rain pelted me. As I peered up at the sky, black swollen clouds completely covered the moon. I stepped back under the fir tree's protective canopy where even ardent rain could not penetrate its protective branches. I sat down on the dry needles with my back supported by the trunk of the tree. I decided to stay awhile until the rain lightened.

The methodical sound of rain falling worked its magic, and I soon became drowsy. Gathering up dry needles as a pillow to rest my head, I lay face up looking up at the upper boughs of the great tree, immovable even in this storm. Its upper canopy was as grand as any cathedral. I was soon fast asleep.

I tend to have prolific dreams in the wee morning hours between 2-4 am. This morning was no exception. I dreamed that I was once again with Coyote. Yet this time, Coyote was standing upright on two legs. He spoke in a language that I understood in the dream, though I do not recall the language now. The rain had stopped, and the silence filled the woods. Coyote's words rose from the silence, like lonely witnesses to existence, in a place no one has ever been to or heard of. The words seem to fall on me, like leaves from a tree or feathers from a bird flying overhead.

Coyote motioned to my white hand with his paw. The hand looked opaque in the ghostly rays of the dream moon. Coyote asked me what it was like to have hands. I raised a bit and basked in the seeming glow of admiration. I spoke proudly about all the wonderful things that hands help human beings do. I spoke about our ability to build things and make things and manipulate things. I spoke for a long time. Coyote watched me as if he was looking past my words to a deep place within me. It seemed like he was sizing me up in a similar manner to when I first encountered him that evening. I got a distinct feeling that I was talking too much. My voice softened and trailed off until it was only a whisper.

The thought passed through my mind that Coyote might be coveting my hands, not admiring them. A tinge of fear quickly swept through my body, up my back into my neck and head. I shyly glanced up at Coyote and saw his open yellow eyes reflecting the moon-glow. Surely, I thought, a being like Coyote without hands, would be jealous, perhaps even resentful, of human hands, hands that have changed the world. I felt a hot blush on my cheeks and stopped speaking as I lowered my gaze. Hoping that he was not able to read my mind.

When I looked back up, I was relieved to see the smiling eyes of Coyote. At first, I was comforted to know there was no danger. A few moments later, I finally understood that this was a ruse. Coyote, the trickster, had caught me, immersed in one of the worst traits of *Homo sapiens*, the egotistical attitude of our "greatness" above all the other species on the planet. A wave of embarrassment swept over me, but this emotion, too, was an indulgence.

Having made his point, Coyote had no intention of wasting any more time. He turned, lowered himself down on all fours, and motioned for me to follow him with his eyes. This time, he ran with full force across the land, his delicate but powerful paws racing over fields, through woods, and across streams, with me trailing close behind. I noticed that for some inexplicable reason, I was able to keep up with him easily now, even at this accelerated pace. I also noticed that I was running just a few feet from the ground and that my eyesight, hearing, and smell were greatly enhanced. There on the Earth passing beneath us, I saw that four paws, not feet, and hands, were transporting me over the land, like a swimmer smoothly cutting through the water. My body just flowed over the terrain. I had been transformed into a Coyote.

We two Coyotes ran through the woods like lightning, our strong legs allowing us to leap over steep ravines and rivers, as our agile bodies avoided boulders and trees in our way. Like the wind, we flew, as if the landscape was created for our feet alone, for our pleasure. When we stopped, which we rarely did, at openings in the trees or on prominences at the tops of open ridges, our throats opened in gratifying howls and yips. In stride, Coyote showed me those secret places revered by the animals that no human being has ever entered.

As we ran, I felt like I had been gifted not only with Coyote's physicality but also with his way of seeing the world and humankind. Our follies, every misguided thought, and destructive act lay before me like a newly opened book of secrets; it almost seemed humorous if our actions had not been so destructive. In many ways, human bodies are less capable of life on Earth than those of other creatures. I saw our vulnerabilities and how our bodies bound us in fears and ignorance. I also understood the compassion, wisdom, and generosity of our kind. Most of all, I saw humankind as simply one of the many life forms on the planet and in the Universe, learning as we go. I was deeply humbled. Soon the sound of laughter rising and then falling away and rising again filled my ears. It sounded both far and near, filling every space surrounding us. I looked side to side but saw no one, and then I understood that it was our laughter alone that rose and fell in unison.

The sound of laughter woke me as I lay on the open meadow, warm, and comfortable in my sleeping bag. The rain had passed, and the dark sky was merging into a smoky blue as dawn arrived. I awoke with an elated joy-filled heart and wondered if I had had a dream of flying or love. As I stretched, a vague memory of running barefoot with Coyote rose in my consciousness. A mist of forgetting clouded my thoughts, and I took this for a dream. I began to plan for my trip home and relished a few more minutes in the warmth of the sleeping bag before getting up.

I reached my hands out searchingly in the dim morning light and felt around for my sneakers. Not finding them, I sat up and looked around the perimeter of my sleeping bag. They were nowhere in sight. I got out of the sleeping bag and stood up. It was not till then that I became very aware of having sore feet and hands. There was dirt between my toes and under my toenails and hand nails. I brushed this off as best I could and began walking around, looking for my sneakers. I found them about twenty minutes later at the edge of the woods. They were side by side as if placed there carefully. A few feet further, on a patch of barren ground leading from the woods, were the paw prints of a coyote, overlaid by the footprints of naked human feet.

My thoughts went back over the dream with Coyote. My heart jumped a bit, and a wave of anxiety and then exhilaration washed through my body. I went and sat back down on the sleeping bag, trying to piece together the activities of the night. It was then that I got a distinct impression that I was being watched. I looked back over the meadow towards the edge of the forest and saw Coyote. He was lying with his haunches and belly on the ground, and his front paws stretched in front. His head was raised, and his yellow eyes looked my way intently. I no longer cared if it was a dream or occurred while I was awake; I just knew it was true.

For an instant, I saw myself through his eyes. Coyote saw me the way that the Creator sees all of us, wild and free. I have never felt so loved. Coyote, the wise trickster, had imparted yet another valuable lesson. Of all the wild animals, Coyote is one who has stayed with us, even as malls, parking lots, and buildings have taken away the wild places. His presence is a reminder of our folly, and it is also a gift. As we over-domesticate and abuse the Earth, wrongly bending it to our dominion, "He Who Cannot Be Domesticated," teaches that there are far greater things in this world. Those who listen to the Earth and learn from the animals and trees will discover these precious treasures for the soul.

28. Becoming A "Wild" Horse

There is no man that shall catch thee by a burst of speed; neither pass thee by, nay, not though in pursuit he were driving goodly Arion, the swift horse of Adrastus, that was of heavenly stock.

Homer, Iliad 23-346

One Spring morning in 2020, a live news feed popped on the computer screen, showing the roundup of a herd of mustangs on Bureau of Land Management (BLM) land. The wild horses are considered pests by BLM because they eat the grasses and drink the water that the Bureau considers the property of ranchers and their domestic cattle.[1] Today, 270 million acres of public land in the 11 western states is used for grazing cattle.[2] Much of the Western United States plains are severely degraded and damaged by over 2 million cattle. (This total does not include cattle owned by "trespass ranchers" like Cliven Bundy, who use public lands illegally without paying a dime, and even protect their trespass with guns and refused to leave.)

Cattle aren't native to this country—they come from Europe, a wetter, greener, and more resilient landscape for cattle hooves. Domestic cattle have also been bred for generations to make the largest steaks. Their abnormally large sizes and bulk, along with their large hooves, tend to destroy native vegetation, damage the land, and muddy and pollute streams, rivers, lakes, and ponds with their waste.[3] Native animals that used to be prevalent in these formerly pristine habitats like wolves and bears, were driven to extinction by tax-supported, government-led "predator control" programs.[4]

When ranchers run their cattle on federally owned lands, they pay a tiny portion of the fair market rental value or none at all. Powerful lobbying groups make sure that this "giveaway" is protected. Ranchers pay just $1.35 a month to graze cattle on public lands and national forests. This tiny fee covers thirty days of grazing on lands held in the public interest and supported by taxpayers' dollars. It is a fee collected by the Bureau of Land Management. Yet, it is not enough to cover the costs to administer the program and certainly not environmental degradation. Thus, US tax dollars are taken to foot the bill to the tune of twenty-five million dollars a year to subsidize ranchers who produce less than 2.7 % of livestock products for the public.

Due to the harm that cattle grazing causes, many other taxpayer supported programs are needed to correct soil erosion, stop desertification and the spread of noxious weeds, and prevent damage to streams, lakes, watersheds, and riparian areas. These programs also have to address the harm done to archeological and cultural resources that belong to native Americans. Even more, money is spent to remedy harm to wildlife caused by cattle ranching. Yet, wildlife is rarely helped; in most cases, tax dollars are used to kill or remove them from "cattle lands.[5]" The USDA spends close to ten million a year to kill millions of native animals "deemed predators" by the ranchers. BLM's wild horse and burro program remove thousands of federally protected horses and burros to provide more forage for domestic cattle to a tune of 80 million.

The entire livestock sector globally is one of the largest natural resources users, such as water and land, and is a major greenhouse gas source. It is the leading cause of water pollution and overconsumption and a primary driver of biodiversity loss. Almost eighty percent of deforestation of the Amazon rainforest is due to making room for cattle and livestock grazing.

When one considers the enormous financial costs and the immeasurable harm to biodiversity and land caused by cattle grazing, it's impossible to justify it! According to the Western Watershed Project,

"Public lands ranching is the most widespread commercial use of public lands in the United States. Ranching is one of the primary causes of native species endangerment in the American West; it is also the most significant cause of non-point source water pollution and desertification. Public lands ranching significantly contributes to climate change by emissions of the global warming gases nitrous oxide and methane; it causes loss of soil carbon reserves by causing erosion and by substantially reducing the landscape's potential to sequester carbon.[6]*"*

~

In the case of mustangs, though they do not damage the land or pollute the water like cattle, they are blocked by BLM from entering "cattle territory", or worse, aggressively "rounded-up."

This morning, the BLM mustang roundup was taking place when the desert flowers were blooming. Shiny mandarin-orange leaves of the Nevada Columbia lily decorated the fertile ground between bitterbrush, desert succulent shrubs, sagebrush, and mixed chaparral. The black stallion leader led a small herd of twenty or so mustang over the dry land for the run of their lives.

Their coats were glistening in the sun from sweat. The stallion and his beloved herd of mares and young colts were being "driven' by the helicopter towards fences and pens created for this purpose. The black stallion's large chest was heaving and dripping wet after running for too long at maximum speed, driven by the loud and frightening whirring of BLM helicopters used to chase the animals.

To the BLM and moneyed lobbyists representing ranchers, wild mustangs do not engender admiration but are simply pests. BLM uses this inhumane roundup method that causes terror and injury to the horses because its quickest, even though many other less injurious and lethal methods are available. Once in the pen, some of the horses that had never been imprisoned in their lives threw themselves against the wire and wood, causing injuries, while others whinnied in a frenzy of fear.

My heart sank at seeing so many beautiful animals being treated so brutally. Then an amazing thing happened. Although near death from exhaustion, and after pacing to and fro within the enclosure, the stallion broke for it. He leaped up and over the fence in one great stride, kicking his heels up so that they just barely made it past the top wire. The cameras' last view was of his black head and mane, held high and shining in the sun, as he disappeared to the other side of a ridge. I stood up and cheered.

~

It has taken millions of years for Eohippus to evolve into the mighty, steadfast, and loyal horse. While the ancestor of the horse, Eohippus, originated in North America 35-56 million years ago and evolved into at least twelve different species, it was Equus that went on to survive and outlast all the rest. During the Pleistocene epoch, 12,000 to 2.5 million years ago, global cooling and humans' arrival led to the extinction of many large mammals. Some horses left their home continent and crossed the Bering land bridge around the beginning of the epoch. Some made it as far as Africa to evolve into zebras. Others arrived in Asia, the Middle East, and northern Africa and evolved into desert friendly species, the wild asses, and onagers.

According to the BLM, ranchers, and others who seek to control and abuse wild horses, all North American horses were wiped out in the Pleistocene. The only remaining horses, the mustangs, were only domestic horses brought to America by the Spaniards and released. This "story" totally disregards Indigenous beliefs, archeological evidence, and oral history. The Native wild horse never went extinct in its homeland. The horse survived the Ice Age and has lived among Native people long before the arrival of the Europeans and their domestic animals.

There is a great deal of archeological and historical evidence that supports this conclusion. Accounts by early explorers refer to Native peoples having horses long before they arrived on the scene. For example, Sir Francis Drake accounts that he came across "herds of horses" living among Native people in California and Oregon. Similarly, the explorer Don Juan de Onate, wrote that New Mexico was full of "wild mares." The first European west of the Hudson Bay, Antony Henday, describes the local Blackfoot as possessing large numbers of native horses in 1754.[7] Besides, there are many archaeological objects and ancient art that depicts horses during the time period that they were supposedly extinct. This includes clay and wood figurines as well as ancient petroglyphs.[8]

Yet, even if horses on the American continent went extinct (which they did not, as ample evidence shows), they are still the same species that originated on the American continent. According to mitochondrial-DNA analysis, the modern or caballine horse, E. caballus, is genetically equivalent to its ancient ancestors that evolved on the American continent.[9] The mustang are indeed the direct descendants of the ancient Native wild horses. It is, therefore, totally irrelevant that some of them have been caught and been domesticated! The BLM and ranchers may find it convenient to refer to wild mustangs as "feral" and "domesticated" so that they can have greater leeway in their control and avoid the types of protections "wild" animals might receive, but their reasoning rests on shaky grounds. Mustangs are the same species that originated in America, and whether or not they were domesticated is irrelevant. Domestication does not alter biology, and it has been proven that when "domesticated" horses are turned loose, they revert to a wild state.[10]

Apart from the sound reasons mentioned above for treating native American horses as protected wild species, not as "feral animals," the horse is one of the only large mammals that has allowed us to domesticate them and use them for our benefit. Without the horse, humankind would never have come as far as we have come today. It is more than patently unfair to treat this worthy animal so inhumanly, and to steal its freedom is immoral. Some would say that it breaks a sacred bond.

According to Yvette Running Horse Collin, a founder of Sacred Way Sanctuary, the horse is not a domesticated "beast of burden" but a *"holy relative gifted to them for caretaking by the Creator long ago... As Indigenous Elders and individuals shared their traditional knowledge and/or their creation stories, it became clear that their perspective regarding the "horse" was almost completely different than that of the dominant Western culture. To their ancestors, these "beings" were "sacred relatives," rather than beasts of burden that existed to serve the whims of humanity. Therefore, the meaning of the words developed by the Indigenous Peoples of the Americas to address this creature are so different from those utilized by the dominant culture, that the word "horse" is actually not even a correct or compatible translation.*[11]"

Human consciousness is intricately connected to the horse's consciousness, when we steal the horse's freedom, we pilfer our own liberty and demean our sacred relationship with the Creator. Many cultures around the world emphasize the sacred dimensions of the horse. For example, the Hindu's have their horse-headed god Hayagriva, revered as the god of wisdom and knowledge. In Norse mythology, it is the greatest of all horses. Sleipnir can fly in the skies, taking Odin to all Nine Worlds. Then, there is the winged stallion Pegasus, the offspring of the Greek God's Poseidon and Medusa. Zeus is said to have welcomed this magnificent flying horse to be with him in the heavens and carry his thunderbolts.

Zeus set Pegasus in the sky as a constellation of stars to last for eternity. This constellation lies in the northern hemisphere and is one of the largest in the sky. The Greek astronomer Ptolemy was the first to describe it residing in the sky between the constellations Pisces and Aquarius. Thus, Pegasus, the horse with wings, carries us from the Piscean Era into the Aquarius's Age with the winds of celestial forces.

Pegasus, the winged horse, is considered special by Zeus because it lives in two realms, the Sky and the Earth. This is humanity's story, as well. We, too, exist in two realms, the manifest world of our daily material lives and the invisible world of spirit, magic, and mystery. As humanity moves into this new Great Age, we must embrace our spirituality and our connection to the invisible while still keeping our feet firmly planted on the sacred ground.

The argument about whether or not the horse is descended from a wild or domestic stock does not take away from the fact that the mustang lives just like a wild horse, facing the same risks and experiencing the same freedoms. It is brave, and it is swift. It is endowed with everything it needs to live and thrive on the plains, where the Sky meets the Earth.

We, too, are like the horse. There may be confusion or disagreements about whether we are wild or tame, yet no matter how far we sink into a state of "domestication", we can always revert and adapt to our natural state. When we are ready, we can be carried ever onwards to freedom by humanity's ally, the horse.

~

As I read the headlines today about a new oil rig being inserted dangerously into the deep ocean, or listen to the tough or misleading talk of those actively destroying the planet while criminally denying that any harm is being done, or see pictures of trophy hunters sitting next to a once-magnificent animal whose life was taken for one person's foolish pride, I cling to this hope: this can change, we can change, this all can change. It's called adaptation, and it has been going on for millions of years.

Perhaps this will occur because someone cares about the future and the seventh generation. Or perhaps someone has a heart-opening exchange like Dr. Pelie, with a magnificent animal or tree. Every person, irrespective of heritage, nationality, or religious belief, has qualities of compassion and the potential to understand and experience all life as sacred. This understanding needs to be put at the forefront of our lives, planted in every leader's heart and shared with the world. Through interaction with the natural world and life itself, there is always the chance that the human heart will open and that the human soul will seek freedom over all else.

~

Doubt not that your intentions and prayers for the Earth will be heard. Or that you have the ability to change the world with your actions.

This is what I wish for you.

In the end, you become the Runaway Horse, with your legs rising high against your belly, lifting you aloft like a winged being, and bearing you breathlessly forward to freedom and joy.

Then, somewhere in the faraway future, one of your descendants will recollect your freedom ride and celebrate your wild gift to them, to life, and to the future.

It is the gift of a "real person."

I will cheer for you then.

~

I still recall today, more than a decade after my flight above the Rio Grande, Jimmy's eyes when speaking of the sanctuaries, the community, and the animals in his care. His kind expression moved me then, and it still does. Because of people like him and so many others, I have hope for the future. One day, descendants of a kind and wise people, who know that humankind's destiny is woven inextricably with the animals, the plants, and the Earth, will remember their ancestors who cared for and inspired a new generation.

In the end, though, it is still Nature herself who comforts and inspires me the most. She is my maker, my truth, my world.

When despair for the world grows in me
and I wake in the night at the least sound
in fear of what my life and my children's lives may be,
I go and lie down where the wood drake
rests in his beauty on the water, and the great heron feeds.
I come into the peace of wild things
who do not tax their lives with forethought of grief.
I come into the presence of still water.
And I feel above me the day-blind stars
waiting with their light. For a time
I rest in the grace of the world and am free.[12]

About the Author

"Nighean Tullimentan"
(Star Shining Daughter)

Catriona (Gaelic for Catherine) is a spiritual teacher, conservationist, scientist and Nature mystic who seeks to heal the broken relationship between humankind and Nature. To learn more about Nature Quests & Celtic Spirit & Nature Services go to catrionamacgregor.org.

She is a visionary bridge builder between nature and humankind—and a natural intuitive. She advises on ways that we can partner with nature to create a healthier and more harmonious world.

Catriona is a descendant of the indigenous people of Scotland. She has been called a shaman, a Celtic Mystic, and even "Mother Nature" for her deep understanding of the natural world.

Catriona managed one of the largest coastal sanctuaries in the US—the gulf coast sanctuary stretching over 600 miles with wintering grounds & stop over sites for 98% of the long distance migratory bird species in N. America. Her program was awarded the Blue Ribbon in Conservation by the Governor and led, in part, to the comeback of endangered species.

Along with her conservation work, she has been called to serve as a 'Spiritual Guide" to help others experience the sacred of the wilds, to connect to their own inner sacred wildness. She is the founder of Nature Quest and has led Vision Quests, Nature Quests, Nature Tours and Spiritual Retreats for two decades.

She is the award-winning author of *Partnering with Nature: The Wild Path to Reconnecting with the Earth* which won a Gold Medal from the Nautilus Book Awards. Previous winners include His Holiness The Dalai Lama, Deepak Chopra, and Thich Nhat Hanh.

Catriona lives in northern California with her husband and house full of animals domestic and wild.

Author at 12 years with sparrow hawk

Notes

Introduction

1. The Picts so named by the Romans were the indigenous people of Scotland and were referred to by those who came later as "Pecht," meaning the elders. They forged a union with the Gaels to create "Alba" Scotland. Today, many of the habits, customs and beliefs of the Celts come from the older society of the Picts. For example, the Picts had Druids and a great deal of astronomical, scientific and architectural knowledge and were great artisans.

2. "European Mistletoe," National Center for Complementary and Integrative Health, U.S. Department of Health and Human Services, last modified September 2016, https://www.nccih.nih.gov/health/european-mistletoe.

3. Terry Robson, An Introduction to Complementary Medicine (Crows Nest, Australia: Allen & Unwin, 2003), 174.

4. Intergovernmental Panel on Climate Change, "Summary for Policymakers," in Climate Change and Land (New York: The United Nations, 2020).

Preface

1. William Wordsworth, "The World is Too Much with Us," Poetry Foundation, accessed July 15, 2020, https://www.poetryfoundation.org/poems/45564/the-world-is-too-much-with-us.

Chapter 1: The Flight

1. Merle Patchett, "On Necro-Ornithologies," Antennae: The Journal of Art and Nature (Spring 2012): 19.

2. "Setting the Scene: Women of the 1890s," Klondike Gold Rush National Historical Park, U.S. National Park Service, last modified April 14, 2015, https://www.nps.gov/klgo/learn/historyculture/women1890s.htm.

3. United States Environmental Protection Agency, "DDT—A Brief History and Status," EPA.gov, last modified August 11, 2017.

4. "The Wright Brothers," PBS.org, accessed June 6, 2020, https://www.pbs.org/wgbh/americanexperience/features/wright-brothers/.

5. Tim Radford, "James Lovelock at 100: the Gaia saga continues," Nature, June 25, 2019.

6. "Kirlian Photography and the Aura" , accessed July 22, 2020, https://web2.ph.utexas.edu/~coker2/index.files/kirlian.shtml,

Chapter 2: How the Animals Shape Us

1. Brave Buffalo, "Dreams Concerning Animals," in Teton Sioux Music, ed. Frances Densmore (Washington: Government Printing Office, 1918), 61: 172, retrieved from https://play.google.com/store/books/details?id=ykUSAAAAYAAJ&rdid=book-ykUSAAAAYAAJ&rdot=1.

2. "The Basics of Bird Migration: How, Why, and Where," All About Birds, January 1, 2007, The Cornell Lab of Ornithology, https://www.allaboutbirds.org/news/the-basics-how-why-and-where-of-bird-migration/.

3. Clifford E. Shackelford et al., Migration and the Migratory Birds of Texas: Who They Are and Where They Are Going (Austin, Texas: Texas Parks and Wildlife, 2005), 2, https://tpwd.texas.gov/publications/pwdpubs/media/pwd_bk_w7000_0511.pdf.

4. Mason Inman, "World's Longest Migration Found—2 Times Longer Than Thought," National Geographic News, January 12, 2010, https://www.nationalgeographic.com/news/2010/1/100111-worlds-longest-migration-arctic-tern-bird/.

5. William K. Purves, David Sadava, and Gordon H. Orians, Life: The Science of Biology: Volume III: Plants and Animals (Sunderland, Massachusetts: Sinauer Associates, 2004), 882.

6. Anonymous, "The Seafarer," trans. Ezra Pound, in The School Bag, eds. Seamus Heaney and Ted Hughes (Faber and Faber, 1997), unpaginated, https://voetica.com/voetica.php?collection=4&poet=743&poem=5246.

7. Job 12:7 (KJV), Bible Gateway, accessed May 29, 2020, https://www.biblegateway.com/passage/?search=Job+12&version=KJV.

8. Plenty-Coups, "Plenty-Coups, Chief of the Crows," ed. Frank B. Linderman, in Native American Autobiography: An Anthology, ed. Arnold Krupat (Madison, Wisconsin: University of Wisconsin press, 1994), 251-253.

9. Arnold Krupat, editor, Native American Autobiography: An Anthology (Madison: University of Wisconsin Press, 1994) 250–253.

10. "Crow Nation," New World Encyclopedia Online, accessed June 23, 2020, https://www.newworldencyclopedia.org/entry/Crow_Nation.

11. Miranda Green, Celtic Goddesses: Warriors, Virgins, and Mothers (New York: George Braziller Press, 1995), 32.

12. "Jataka Tale: Prince Mahasattva," Dunhuang Foundation, May 7, 2018, http://dunhuangfoundation.us/blog/2018/3/7/jataka-tale-prince-mahasattva.

13. Roseh Dalal, Hinduism: An Alphabetical Guide (London: Penguin, 2010), 112.

14. Richard H. Wilkinson, The Complete Gods and Goddesses of Ancient Egypt (London: Thames and Hudson, 2003), 202.

15. Richard Wilkinson, The Complete Gods and Goddesses of Ancient Egypt. Thames & Hudson, 2003.

16. Thomas Green, Concepts of Arthur (Stroud, United Kingdom: Tempus, 2007), 259, 261-262.

17. "Tower of London: The ravens," Historic Royal Palaces, accessed June 23, 2020, https://www.hrp.org.uk/tower-of-london/whats-on/the-ravens/.

18. Linda Tucker, Mystery of The White Lions: Children of the Sun God (Carlsbad, California: Hay House, 2010), 64.

19. Carl Jung, Modern Man in Search of a Soul, trans. Cary F. Baynes and William Stanley Dell (London: Kegan Paul, Trench, Trubner and Co., 1933), 67.

Chapter 3: Pelican

1. "Holy Eucharist Teachings Index," The Work of God, accessed April 29, 2020, https://www.theworkofgod.org/Devotns/Euchrist/holy_eucharist.asp?psearch=Aquinas.

2. Alan Feduccia, The Origin and Evolution of Birds (New Haven, Connecticut: Yale University Press, 1999), 187.

3. Bob Strauss, "Prehistoric Life During the Eocene Epoch," ThoughtCo., last modified August 16, 2019, https://www.thoughtco.com/the-eocene-epoch-1091365.

4. Bruce J. MacFadden, "Fossil Horses—Evidence for Evolution," Science 307, no. 5716 (March 2005): 1728-1730.

5. Matt Walker, "Oldest prehistoric pelican also had big beak," Earth News, last modified June 11, 2010, http://news.bbc.co.uk/earth/hi/earth_news/newsid_8733000/8733503.stm.

6. Abdallah Diab, "From an Immigrated Bird to a Deity: Pelican in Ancient Egyptian Sources," International Journal of Heritage, Tourism, and Hospitality 11 (2017): 87-95.

7. Richard Stracke, "The Pelican Symbol," A Guide to Christian Iconography: Images, Symbols, and Texts, 2018, https://www.christianiconography.info/pelicans.html.

8. "Elizabeth I's pelican emblem," Royal Museums Greenwich, accessed July 7, 2020, https://www.rmg.co.uk/explore/elizabeth-pelican-emblem#:~:text=The%20pelican&text=This%20potent%20symbol%20of%20self,of%20Queen%20Elizabeth's%20favourite%20symbols.

9. Thomas Aquinas, "Adoro te devote," in "Adoro Te Devote by St. Thomas Aquinas," Catholic Online, accessed May 25, 2020, https://www.catholic.org/prayers/prayer.php?p=2196.

10. "The Clock of Eras and Geologic Time," Fossils, Facts and Finds, accessed June 7, 2020, https://www.fossils-facts-and-finds.com/clock_of_eras.html#:~:text=The%20Clock%20of%20Eras%20uses,our%20planet%20in%20geologic%20time.&text=The%20Clock%20represents%20geologic%20time,represents%20approximately%20375%20million%20years.

11. Jerry Sullivan, "The Passenger Pigeon: Once There Were Billions," in Hunting for Frogs on Elston, and Other Tales from Field and Street, ed. Victor M. Cassidy (Chicago: University of Chicago Press, 2004), 210-213.

12. Ignacio Pujana and Robert Stern, "U.S. Mexico Water Resources: The Rio Grande/Rio Bravo example," in Geology, Resources, and Environment of Latin America.

Chapter 4: Eiocha

1. *Oran Mór*, as quoted in Orna Ross, "A Celtic Creation Story," Orna Ross: Indie Novelist and Poet, April 7, 2016, https://www.ornaross.com/a-celtic-creation-story/.

2. "Pinto Horse," International Museum of the Horse, accessed June 23, 2020, http://imh.org/exhibits/online/breeds-of-the-world/north-america/pinto-horse/.

3. Bonnie Hendricks, *International Encyclopedia of Horse Breeds* (Norman, Oklahoma: University of Oklahoma Press, 2007), 35.

4. Hendricks, *International Encyclopedia of Horse Breeds*, 35.

5. Deb Bennett, *Conquerors: The Roots of New World Horsemanship* (Lubbock, Texas: Amigo Publications, 1998).

6. CuChullaine O'Reilly, "The History of Equestrian Travel," The Long Riders' Guild, accessed June 25, 2020, http://www.thelongridersguild.com/stories/history-of-ET-2000.htm.

7. Joshua Hammer, "Finally, the Beauty of France's Chauvet Cave Makes its Grand Public Debut, *Smithosian Magazine*, April 2015, https://www.smithsonianmag.com/history/france-chauvet-cave-makes-grand-debut-180954582/.

8. Mariko Namba Walter and Eva Jane Neumann Fridman, eds., "Horses," in *Shamanism: An Encyclopedia of World Beliefs, Practices, and Culture* (Santa Barbara, California; Denver, Colorado; Oxford, England: ABC-CLIO), 1:148.

Chapter 5: Selkie

1. Meeting with the Kogi. Private Estate Northern California.

2. Katie Morosky, "Rare dolphin sighting in Long Island Sound: Unusually large group of dolphins spotted near Riverhead," *RiverheadLOCAL*, August 17, 2015.

3. Ben Panko, "Forget Dinos: Horseshoe Crabs Are Stranger, More Ancient—And Still Alive Today," *Smithsonian Magazine*, July 5, 2017.

4. W. H. Hudson, Green Mansions: A Romance of the Tropical Forest.1904, page 35,37.

5. Hudson, Green Mansions , Pge. 63

6. W. Y. Evans-Wentz, *The Fairy-Faith in Celtic Countries* (New York: Citadel, 1990), 167, 243, 457.

7. Evans-Wentz, *The Fairy-Faith in Celtic Countries*, xx.

8. Robert Kirk, The Secret Commonwealth of Elves, Fauns and Fairies, Published by David Nutt, In the Strand, London, 1893, pg 74.

9. Masaru Emoto, *The Hidden Messages in Water* (New York: Atria Books, 2011).

10. Whitney Clavin and Alan Buis, "Astronomers Find Largest, Most Distant Reservoir of Water," NASA.gov, July 22, 2011.

11. Krishna Dvaipayana Vyasa, *The Mahabharata of Krishna-Dvaipayana Vyasa Translated into English Prose*, trans. Kisari Mohan Ganguli (Calcutta, India: Bharata Press, 1883-1896).

12. "What it means to give the Whanganui River the same rights as a person," *Radio New Zealand*, March 16, 2017, https://www.rnz.co.nz/news/the-wireless/374515/what-it-means-to-give-the-whanganui-river-the-same-rights-as-a-person.

Chapter 6: Last Child in the Woods

1. Walt Whitman, "There Was a Child Went Forth," in *Leaves of Grass: The First (1855) Edition*, ed. Malcolm Cowley (London: Penguin Books, 1959), 138.

2. R. Brian Ferguson, "War Is *Not* Part of Human Nature," *Scientific American*, September 1, 2018.

3. Margaret Mead, "Warfare is Only an Invention—Not a Biological Necessity," in *The Dolphin Reader*, ed. Douglas Hunt (Boston: Houghton Mifflin Company, 1990), 415-421.

4. "Kirlian Photography," in *Funk and Wagnalls New World Encyclopedia* (New York: Funk and Wagnalls Co., 2010).

5. Edwin Hall and Horst Uhr, "Aureola Super Auream: Crowns and Related Symbols of Special Distinction for Saints in Late Gothic and Renaissance Iconography," *The Art Bulletin* 67, no. 4 (December 1985): 567-603.

6. The Editors of Encyclopaedia Britannica, "Chakra," Encyclopaedia Britannica Online, last updated October 10, 2018.

7. Michael Hedelberger, *Nature from Within: Gustav Theodor Fechner and His Psychophysical Worldview*, trans. Cynthia Klohr (Pittsburgh, Pennsylvania: University of Pittsburgh Press, 2004).

8. Liliana Usvat, "Kirlian Photography, Trees Perception of Human Intention, Biocommunication Aura," Environment and Society, November 18, 2013, https://lilianausvat.wordpress.com/2013/11/18/kirilian-photography-trees-perception-of-human-intention-biocommunication-aura/.

9. Peter Tompkins and Christopher Bird, *The Secret Life of Plants* (New York: Harper and Row, 1973), 10.

10. Tompkins and Bird, *The Secret Life of Plants*, 11.

11. Tompkins and Bird, *The Secret Life of Plants*, 7.

12. Yin et al., "Satellite-based entanglement distribution over 1200 kilometers," *Science* 365, no. 6343 (June 2017): 1140-1144.

13. Time-Life Books Editors, *Earth Energies* (Alexandria, VA: Time-Life Books, 1991), 132–133. Tompkins and Bird, *The Secret Life of Plants*, 10.

14. Jurriaan Kamp, "They're All Ears," *Ode Magazine*, December 2007, 17.

15. Jihye Jung, et al, "Beyond Chemical Triggers: Evidence for Sound-Evoked Physiological Reactions in Plants", Frontiers of Plant Science. 2018. Published online 1.30.2018. 10.3389/fpls.2018.00025

16. Ankur Patel, Sangeetha Shankar, Seema Narkhede, "Effect of Sound on the Growth of Plants, ResearchGate 2016. https://www.researchgate.net/publication/313903754_Effect_of_Sound_on_the_Growth_of_Plant_Plants_Pick_Up_the_Vibrations/citation/download, accessed on July 22, 2020.

17. Sylvia Hughes, "Antelope Activate the Acacia's Alarm System", New Scientist, 9.29. 1990.

18. Sara Lasgow, The Hidden Memories of Plants: Inside a quiet revolution in the study of the world's other great kingdom, AtlasObscura, SEPTEMBER 5, 2017 https://www.sarahlaskow.com/stories/2018/5/21/the-hidden-memories-of-plants

19. Peter Wohlleben, *The Secret Life of Trees* (Vancouver: Greystone Books, 2016).

20. Rupert Sheldrake, *A New Science of Life: The Hypothesis of Formative Causation* (London: Icon Books, 2009).

21. In that case, a group of rats were trained in how to escape from a water maze. It was found that not only were later generations able to escape the maze more easily than their parents or ancestors, but that unrelated rats used in other countries after this experiment appeared to have somehow "absorbed" this learning as well. (See W.E. Agar, F.H. Drummond, O.W. Tiegs, MM. Gunson, "Fourth (final) report on a test of McDougall's Lamarckian experiment on the training of rats," *Journal of Experimental Biology* 31, no. 3 (September 1954): 307-321.)

22. Laird Scranton, "Revisiting Griaule's Dogon Cosmology." *Anthropology News*, Vol. 48, No 4 (April 2007)

23. The author of four books, including *Light After Life: Experiments and Ideas on After-Death Changes* (1998, NY, Backbone Publishing Co.), and over 70 articles in leading journals on physics and biology, he holds twelve patents on biophysics inventions.

24. Hornung, Erik; David Lorton (15 June 1999). *The ancient Egyptian books of the afterlife*. Cornell University Press. p. 14. ISBN 978-0-8014-8515-2.

Chapter 7: The Picts: People of the Land and Sea

1. "Picts," in *Collins Encyclopedia of Scotland*, eds. John Keay and Julia Keay (London: Trafalgar Square, 2001), 755.

2. They were referred to as "Pecht", meaning ancient ones. The Romans likely misinterpreted this to mean "Pict" and some people have interpreted that to mean painted. However, while the Picts may have worn temporary tattoos in times of battle there are no depictions of painted people in Pictish carvings or art.

3. Kathryn Krakowka, "Prehistoric pop culture: Deciphering the DNA of the Bell Beaker Complex," *Current Archeology*, April 5, 2018.

4. Sigurd Towrie, "Viking Orkney," Orkneyjar: The Heritage of the Orkney Islands, accessed July 16, 2020, http://www.orkneyjar.com/history/vikingorkney/takeover.htm.

5. Adam Hart-Davis, "Hadrian's Wall: A horde of ancient treasures make for a compelling new Cumbrian exhibition," *The Independent*, June 23, 2011.

6. Julius Caesar, *The Gallic Wars*, trans. W.A. McDevitte and W.S. Bohn (London: G. Bell and Sons, 1927), E-Book, Book 4, Chapter 33, http://classics.mit.edu/Caesar/gallic.4.4.html.

7. "Celts and Scythians Linked by Archaeological Discoveries," Beyond Today, accessed March 6, 2020, https://www.ucg.org/bible-study-tools/booklets/the-united-states-and-britain-in-bible-prophecy/celts-and-scythians-linked-by-archaeological-discoveries.

8. "The Geography of Celtic-Scythian Commerce," Beyond Today, accessed March 6, 2020, https://www.ucg.org/bible-study-tools/booklets/the-united-states-and-britain-in-bible-prophecy/the-geography-of-celtic-scythian-commerce.

9. Iñigo Olalde et al., "The Beaker phenomenon and the genomic transformation of northwest Europe," *Nature* 555 (2018): 190-196.

10. "Epidii," Oxford Reference, Oxford University Press, accessed July 17, 2020, https://www.oxfordreference.com/view/10.1093/oi/authority.20110810104839316.

11. Joshua J. Mark, "Tacitus' Account of the Battle of Mons Graupis," Ancient History Encyclopedia, January 9, 2015.

12. Joshua J. Mark, "Picts," Ancient History Encyclopedia, December 18, 2014, Ancient History Encyclopedia Foundation, https://www.ancient.eu/picts/.

13. Sally M. Foster, *Picts, Gaels, and Scots: Early Historic Scotland* (London: Batsford, 1996), 26-28.

14. Bruce W. Frier and Thomas A.J. McGinn, *A Casebook on Roman Family Law* (Oxford: Oxford University Press, 2004), 19-20.

15. Sarah Pruitt, "Who Was Boudica?" History.com, last updated March 7, 2019, https://www.history.com/news/who-was-boudica.

16. *The Annals of Tigernach*, ed./trans. Stokes Whitley, (Paris: Librairie Émile Bouillon, 1895), 390.

17. The Editors of the Encyclopaedia Britannica, "Boudicca," Encyclopaedia Britannica, Encyclopaedia Britannica, Inc, last modified March 25, 2020, https://www.britannica.com/biography/Boudicca#ref210468.

18. The Editors of the Encyclopaedia Britannica, "Boudicca."

19. Stephanie Lawson, "Nationalism and biographical transformation: The case of Boudicca," *Humanities Research* 19, no. 1 (2013): 108-109.

20. "The 200-year-old Orkney festival where girls dress as horses," *The Scotsman*, March 17, 2016, https://www.scotsman.com/whats-on/arts-and-entertainment/200-year-old-orkney-festival-where-girls-dress-horses-1480316.

21. Edwin Muir, "The Horses," in *Edwin Muir, Selected Poems*, T.S. Eliot, Editor (Faber & Faber Ltd., London, United Kingdom: 1965).

Chapter 8: Women and Nature

1. Anthon, Charles (1855). "Artemis." A Classical dictionary. New York: Harper & Brothers. p. 210.

2. 1 Timothy 2:11-2:14 (NIV), Bible.

3. "Boxed In: Women and Saudi Arabia's Male Guardianship System," Human Rights Watch, 2016, https://www.hrw.org/report/2016/07/16/boxed/women-and-saudi-arabias-male-guardianship-system#.

4. World Economic Forum, "The Global Gender Gap Report," (Cologny, Switzerland: World Economic Forum, 2018), vii.

5. UN Women, "Equal pay for work of equal value," UN Women, accessed June 10, 2020, https://www.unwomen.org/en/news/in-focus/csw61/equal-pay.

6. Monique Villa, "Women own less than 20% of the world's land. It's time to give them equal property rights," January 11, 2017, World Economic Forum, https://www.weforum.org/agenda/2017/01/women-own-less-than-20-of-the-worlds-land-its-time-to-give-them-equal-property-rights/.

7. United Nations Children's Fund, Ending Child Marriage: Progress and prospects, UNICEF, New York, 2014.

8. "A staggering one-in-three women experience physical, sexual abuse," UN News, November 24, 2019, https://news.un.org/en/story/2019/11/1052041.

9. Vandana Shiva, "Women Feed the World, Not Corporations," Zed, September 7, 2016.

10. National Center on Caregiving at Family Caregiver Alliance, "Women and Caregiving: Facts and Figures," Family Caregiver Alliance, December 31, 2003.

11. Bruce W. Frier and Thomas A.J. McGinn, A Casebook on Roman Family Law (Oxford: Oxford University Press, 2004), 19-20.

12. Claudia Moser, "Eastern Religions in the Roman World," Heilbrunn Timeline of Art History, April 2017, The Metropolitan Museum of Art.

13. Pliny the Elder, Natural History, trans. John Bostock and H.T. Riley, vol. 36, chapter 21 (London: Taylor and Frances, 1855).

14. Antipater of Sidon, "On the Temple of Artemis at Ephesus," in Greek Anthology, ed. W.R. Paton (London: William Heinemann Ltd., 1915), Volume 3, Book 9: 58.

15. Gilbert Torres, "The Cult of Artemis and the Royal Priesthood," Round Rock Ministry, January 23, 2014.

16. Anne Soukhanov, ed., "Trivia," in Microsoft Encarta College Dictionary: The First Dictionary for the Internet Age (New York: St. Martin's Press, 2001), 1538.

17. Kevin Mahoney, Latdict: Latin Dictionary and Grammar Resources, accessed June 20, 2020, https://latin-dictionary.net/.

18. E. Cobham Brewer, Dictionary of Phrase and Fable (Philadelphia: Henry Altemus, 1894), 593, 1246.

19. (Pagans is a term that does not do justice to the sophistication of the many cultures that experienced Earth-based divinity such as the Phoenicians and Picts.)

20. Martin Carver, Portmahomack: Monastery of the Picts (Edinburgh: Edinburgh University Press, 2008), 23.

21. Alexander Boyle, "Matrilineal Succession in the Pictish Monarchy," The Scottish Historical Review 56, no. 161 (April 1977): 1-10.

22. Tobias Fischer-Hansen and Birte Poulsen, eds., From Artemis to Diana: The Goddess of Man and Beast (Collegium Hyperboreum and Museum Tusculanum Press, 2009), 245.

23. Aaron J. Astma, "Artemis Estate," Theoi Project, accessed June 24, 2020, https://www.theoi.com/Olympios/ArtemisTreasures.html.

24. Rhona Lewis, "How Many Eggs Are Women Born With? And Other Common Questions About Egg Supply," Healthline, December 18, 2019, https://www.healthline.com/health/womens-health/how-many-eggs-does-a-woman- have#:~:text=Yes%2C%20female%20babies%20are%20born,are%20made%20during%20your%20lifetime.

25. Biological functions do not automatically make women good mothers, and men unable to fulfill this nurturing role. Many women choose to forgo children to further their careers and creative endeavors and many men are wonderful parents. Yet, while many men are involved in childcare, it is still the mother in the majority of families who spends the most time caring for the child or elderly parents. Compared to women, men also tend to remain more resistant to going green. According to Scientific America "Women have long surpassed men in the arena of

environmental action; across age groups and countries, females tend to live a more eco-friendly lifestyle. Compared to men, women litter less, recycle more, and leave a smaller footprint. Some researchers have suggested that personality differences, such as women's prioritization of altruism, may help to explain this gender gap in green behavior" With that said, some men have played critically important roles in protecting the environment. Visionaries like John Muir, David Brower, David Attenborough, and many more have worked tirelessly for this cause.

26. Matrilineal societies not only instill greater rights for women, but they also tend to bring other benefits as well for all people, men, women, and children alike. For example, in a matrilineal society there is no such thing as "illegitimate" children. All children know their empowered mothers, from whom they can inherit property and have a rightful place within their society.

Chapter 9: The Real People

1. Percy Blysshe Shelley, "The Triumph of Life," Poetry Foundation, Poetry Foundation, accessed July 16, 2020, https://www.poetryfoundation.org/poems/45143/the-triumph-of-life.

2. Doug O'Hara, "Coalition Seeks to Protect Bering Sea," *Anchorage Daily News*, August 6, 2003, 6.

Chapter 10: The Rise and Fall of Consumers

1. Mahatma Gandhi, as quoted on GoodReads.com, accessed May 27, 2020, https://www.goodreads.com/quotes/30431-earth-provides-enough-to-satisfy-every-man-s-needs-but-not.

2. E.F. Schumacher, *Small is Beautiful: Economics as if People Mattered* (New York: HarperPerennial, 1989), 15.

3. Heard orally at Bering Sea Summit, story modified by author.

4. Yet, the same taxpayers that subsidized the industry pay full cost for the product (such as gasoline or natural gas) at the pump or pipe. Then taxpayers pay a third time when their tax dollars are used to clean up the pollution that fossil fuel industries cause.

5. United States Congress, *Congressional Record: Proceedings and Debates of the 87th Congress*, 2nd ed., (Washington DC: United States Government Printing Office, 1962), Volume108, Part 14: 18577.

6. "How many hours did people work in history," lovemoney.com, accessed June 12, 2020, https://www.lovemoney.com/gallerylist/84600/how-many-hours-did-people-work-in-history.

7. Phoebe Weston, "Secret to happiness? Having good FRIENDS: 80-year-long Harvard study finds relationships are more important to contentment than money or success," Daily Mail, April 20, 2018, https://www.dailymail.co.uk/sciencetech/article-5639303/80-year-long-Harvard-study-confirms-relationships-important-money-success.html#:~:text=Having%20good%20FRIENDS%3A%20 80%2Dyear,contentment%20than%20money%20or%20 success&text=An%2080%2Dyear%20long%20Harvard,happiness%20th-an%20money%20or%20success.

8. Adam Minter, "Somebody's Making Money Off of All Our Junk," *Bloomberg Business Week*, August 28, 2017.

9. Louise Story, "Anywhere the Eye Can See, It's Likely to See an Ad," *The New York Times*, January 15, 2007.

10. Jeanne E. Arnold et al., *Life at Home in the Twenty-First Century: 32 Families Open Their Doors* (Los Angeles: Cotsen Institute Press, 2012).

11. Associated Press, "Why Muslims view Quran as sacred," *SF Gate*, September 9, 2010.

12. Joey Korn, "Spiritual Space Clearing: Exploring Subtle Energies with Joey Korn," Dowsers.com, accessed March 28, 2020, https://dowsers.com/.

Chapter 11: Goodbye to the Trees

1. William R.L. Anderegg, Jeffrey M. Kane, and Leander D.L. Anderegg, "Consequences of widespread tree mortality triggered by drought and temperature stress," *Nature Climate Change* 3 (2013): 30-36.

2. Jim Robbins, "Bark Beetles Kill Millions of Acres of Trees in West," *The New York Times*, November 17, 2008.

3. Kyle Dickman, "How a wildfire kicked up a 45,000-foot column of flames," *Popular Science*, June 21, 2017.

4. A new kind of imaging spectrometer able to use sunlight to measure and determine the water content of a tree has shown that even green appearing trees may be stressed beyond their limits and dead within the next 6-12 months. According to the spectroscopy, millions of trees are at risk of dying unless weather patterns improve. This is because during a drought the root systems of trees are damaged. They are then less able to access the water they need by way of their roots and are also subject to bending and falling over. The spectrometer is stationed on board the Carnegie Airborne Observatory that is flying over forest lands in California to help predict where the next major die-offs will occur. Even the iconic Ponderosa Pine, Pinus Ponderosa, are thousands of years old and have survived all kinds of environmental challenges are in trouble. Forest Service professionals believe that these trees will all be gone in just 10-20 years.

5. Kurtis Alexander, "Report: California's tree die-off reaches 147 million, boosting fire threat," *San Francisco Chronicle*, February 11, 2019, https://www.sfchronicle.com/california-wildfires/article/Report-Drought-s-end-slowed-California-s-13607328.php.

6. "Forest Research and Outreach," University of California: Agriculture and Natural Resources, accessed July 11, 2020, https://ucanr.edu/sites/forestry/California_forests/.

7. David Siders, "Jerry Brown declares emergency for dying trees," *The Sacramento Bee*, October 30, 2015.

8. Joan Roach, "Source of Half Earth's Oxygen Gets Little Credit, *National Geographic Magazine*, June 7, 2004.

9. Curtis M. Bradley, Chad T. Hanson, Dominick A. DellaSala, "Does Increased Forest Protection Correspond to Higher Fire Severity in frequent forest fires of the Western United States." Ecosphere, Vol 7. Issue 10. 2016.

10. Justin Gillis, "With Deaths of Forests, a Loss of Key Climate Protectors," *The New York Times*, October 1, 2011.

11. Eleanor Ainge Roy, "New Zealand glaciers turn brown from Australian bushfires' smoke, ash and dust," *The Guardian*, January 1, 2020.

12. "It's very analogous to El Niño or 'the blob,' something that's occurring that causes the atmosphere to move around, which causes these warmer or cooler conditions, or wetter and drier conditions, somewhere else," according to the study's author, Abigail Swann, a university professor of atmospheric sciences and biology. (See Hannah Hickey, "Forest loss in one part of US can harm trees on the opposite coast," *UW News*, May 15, 2018, https://www.washington.edu/news/2018/05/15/forest-loss-in-one-part-of-us-can-harm-trees-on-the-opposite-coast/.)

13. Daniel Griffin and Kevin J. Anchukaitis, "How unusual is the 2012-2014 California drought?" *Geophysical Research Letters* 41 (December 2014): 9017-9023.

14. Robin Wylie, "Severe droughts explain the mysterious fall of the Maya," *BBC Earth*, February 22, 2016.

15. Brian Handwerk, "The American West May Be Entering a 'Megadrought' Worse Than Any in Historical Record," *Smithsonian Magazine*, April 16, 2020.

16. A. Park Williams et al., "Large contribution from anthropogenic warming to an emerging North American megadrought," *Science* 368, no. 6488 (April 2020): 314-318.

17. National Audubon Society, "California's Common Birds in Decline," audobon.org, accessed July 11, 2020, https://ca.audubon.org/californias-common-birds-decline.

18. David R. Edmunds Matthew J. Kauffman,Brant A. Schumaker,Frederick G. Lindzey,Walter E. Cook,Terry J. Kreeger,Ronald G. Grogan,Todd E. Cornish. "Chronic Wasting Disease Drives Population Decline of White-Tailed Deer," Plos One, August 30, 2016, https://journals.plos.org/plosone/article?id=10.1371/journal.pone.016112

19. Neela Banerjee, Lisa Song, and David Hasemyer, "Exxon's Own Research Confirmed Fossil Fuels' Role in Global Warming Decades Ago," *Inside Climate News*, September 16, 2015. Geoffrey Supran and Naomi Oreskes, "Assessing ExxonMobil's climate change communications (1977-2014)," *Environmental Research Letters* 12, no. 8 (August 2017).

20. Shannon Hall, "Exxon Knew about Climate Change almost 40 years ago," *Scientific American*, October 26, 2015.

21. Banerjee, Song, and Hasemyer, "Exxon's Own Research Confirmed Fossil Fuels' Role in Global Warming Decades Ago."

22. Oliver Milman and Fiona Harvey, "US is hotbed of climate change denial, major global survey finds," *The Guardian*, May 8, 2019.

23. M. Stuiver, R.L. Burk, P.D. Quay, "13C/12C ratios in tree rings and the transfer of biospheric carbon to the atmosphere," *Journal of Geophysical Research: Atmospheres* 89 (December 1984): 11731-11748.

24. Eric Steig, "How do we know that recent CO2 increases are due to human activities?" RealClimate: Climate science from climate scientists, December 22, 2004, http://www.realclimate.org/index.php/archives/2004/12/how-do-we-know-that-recent-cosub2sub-increases-are-due-to-human-activities-updated/.

25. Charles Choi, Earth's Atmospheric Oxygen Levels Continue Long Slide, Live Science, September 22, 2016, https://www.livescience.com/56219-earth-atmospheric-oxygen-levels-declining.html accessed on February 2020

26. J.E. Harries et al., "Increases in greenhouse forcing inferred from the outgoing longwave radiation spectra of the Earth in 1970 and 1997," *Nature* 410, no. 6828 (2001): 355-357.

27. Kevin Trenberth, "Global Warming is Happening," National Center for Atmospheric Research, accessed March 25, 2020.

28. Peter Rejcek, "Going Deep: Drilling project to retrieve longest ice core ever from South Pole," *In-Depth Newsletter*, National Science Foundation Ice Core Facility, Spring 2015.

29. Shaun A. Marcott et al., "Centennial-scale changes in the global carbon cycle during the last deglaciation," *Nature* 514, nos. 616-619 (October 2014).

30. Morgan F. Schaller, James D. Wright, and Dennis V. Kent, "Atmospheric *P*CO2 Perturbations Associated with the Central Atlantic Magmatic Province," *Science* 331, no. 6023 (March 2011): 1404-1409.

31. Denise Breitburg et al., "Declining oxygen in the global ocean and coastal waters," *Science* 359, no. 6371 (January 2018).

32. Max Roser, "Forests," Our World in Data, Global Change Data Lab, accessed July 17, 2020, https://ourworldindata.org/forests#global-forest-cover-change-over-the-last-centuries.

33. Secretariat of the Convention on Biological Diversity, *Global Diversity Outlook* 3 (Montreal: Secretariat of the Convention on Biological Diversity, 2010), 9.

34. Adam Sandberg, "Eco-Anxiety Takes a Toll on Global Warming Alarmists," Global Warming.org, March 25, 2013, http://www.globalwarming.org/2013/03/25/eco-anxiety-takes-its-toll-on-global-warming-alarmists/.

35. Leslie Crawford, "Green with Worry," *San Francisco Magazine*, February 2008.

36. Samuel Taylor Coleridge, "The Rime of the Ancient Mariner (text of 1834)," Poetry Foundation, accessed March 7, 2020, https://www.poetryfoundation.org/poems/43997/the-rime-of-the-ancient-mariner-text-of-1834.

37. Laurette Rogers and Ginger Potter, "Students and Teachers Restoring A Watershed (STRAW)," eePRO: The hub for environmental education professional development, accessed March 7, 2020, https://naaee.org/eepro/groups/global-ee/case-study/students-and-teachers-restoring.

Chapter 12: A Calling

1. James Allen, "Chapter Six: Visions and Ideals," in As a Man Thinketh, (Manchester, UK: Savoy Publishing Company, 1903).

2. "Chartres Cathedral," Sacred Land Film Project, March 1, 2000, https://sacredland.org/chartres-cathedral-france/.

3. Paul Halsall, ed./trans., "Innocent VIII: BULL Summis desiderantes, Dec. 5th, 1484," Internet Medieval Sourcebook: Witchcraft Documents [15th Century], Fordham University, January 26, 1996.

4. Heinrich Kramer, The Hammer of Witches: A Complete Translation of the Malleus Maleficarum, trans. Cristopher S. Mackay (Cambridge, UK: Cambridge University Press, 2014).

5. Steve Hendrix, ""The Salem witch trials: Why everyone from Trump to Woody Allen still invokes their hysteria," The Washington Post, October 17, 2017.

6. The Burning Times, directed by Donna Read (1990; Toronto, Canada: National Film Board of Canada).

7. Barbara Ehrenreich and Deirdre English, "Witches, midwives, and nurses: A history of women healers," libcom.org, December 27, 2012.

8. William K. Boyd and Henry W. Meikle, eds., "[Confession of Agnes Samsone]," in Calendar State Papers Scotland, vol. 10 (Edinburgh: His Majesty's General Register House, 1936), https://www.british-history.ac.uk/cal-state-papers/scotland/vol10.

9. Leigh Whaley, "The Wise-Woman as Healer: Popular Medicine, Witchcraft and Magic," in Women and the Practice of Medical Care in Early Modern Europe, 1400-1800 (London: Palgrave Macmillan, 2011), 174-195.

10. Read, The Burning Times.

11. Bonnie G. Smith, "HERETICS," in The Oxford Encyclopedia of Women in World History (Oxford: Oxford University Press, 2008), 1: 450.

12. Lyndal Roper, Witch Craze: Women and Evil in Baroque Germany (New Haven, Connecticut: Yale University Press, 2004), 160.

13. Peter Toth, "River Ordeal—Trial by Water—Swimming of Witches: Procedures of Ordeal in Witch Trials," in Witchcraft Mythologies and Persecutions, eds. Gabor Klaniczay and Eva Pocs (Budapest: Central European University Press, 2008), 129-135.

14. Matilda Joslyn Gage, Woman, Church and State: A Historical Account of the Status of Woman Through the Christian Ages: With Reminiscences of the Matriarchate (Chicago: Charles H. Kerr and Company, 1893), 272.

15. Many women would even agree to the "confessions" that were invented by their persecutors just to stop the agony or save their loved ones. All were forced to agree to the most absurd stories. "Kissing the devil's ass" seemed to be a favorite confession forced by Inquisitors. Now you know where the expression of "kiss my ass" came from. In the case of Agnes Sampson, who was pinned to the wall using a witch's bridle until she confessed to over fifty charges against her, the invented confession claimed that she threw parts of a dead cat into the sea to create a storm to shipwreck the King.

16. Anthony Faiola, "8 of Pope Francis's most liberal statements," The Washington Post, September 7, 2015, https://www.washingtonpost.

com/news/worldviews/wp/2015/09/07/what-has-pope-francis-actually-accomplished-heres-a-look-at-7-of-his-most-notable-actions/.

17. "'Last witch in Europe' cleared," SWI swissinfo.ch, August 27, 2008, https://www.swissinfo.ch/eng/-last-witch-in-europe--cleared/662078.

18. Bob Weinhold, "Epigenetics: The Science of Change," Environ Health Perspectives, March 2006, -https://www.ncbi.nlm.nih.gov/pmc/articles/PMC1392256.

19. Olga Khazan, "Inherited Trauma Shapes Your Health," The Atlantic, October 16, 2018.

20. For example, in one study mammals that received poor maternal care did not properly groom themselves when they matured and tended to die earlier. Yet, after giving them the drug trichostatin, this reversed the lack of mothering experience and led to the mature offspring taking better care of themselves.

21. Dawson Church, A Genie in Your Genes: Epigenetic Medicine and the New Biology of Intention, Energy Psychology Press, 2009.

22. Association of American Medical Colleges, "Number and Percentage of Active Physicians by Sex and Specialty, 2017," Washington, D.C.: Association of American Medical Colleges, 2017, https://www.aamc.org/data-reports/workforce/interactive-data/active-physicians-sex-and-specialty-2017.

23. Mathieu Boniol et al., "Gender equity in the health workforce: Analysis of 104 countries. Working paper 1," Geneva: World Health Organization, 2019, https://apps.who.int/iris/bitstream/handle/10665/311314/WHO-HIS-HWF-Gender-WP1-2019.1-eng.pdf?ua=1.

24. Lyn Freeman, "Therapeutic Touch: Healing with Energy," in Mosby's Complementary and Alternative Medicine: A Research-Based Approach, 3rd ed. (St. Louis, Missouri: Mosby Elsevier, 2008), 519-532.

25. Timothy Sellati, Parallel Pandemics: Covid 19 and Lyme Disease, Global Lyme Alliance, April 2, 2020. oballymealliance.org/parallel-pandemics-covid-19-and-lyme-disease/ accessed July 23.

26. Alan MacDonald, Thomas Grier, Paula Pierce, "Diffuse cortical lewy body dementia—two cases-linked by FISH studies of

DNA hybridization and immunohistochemtry to tertiary Borrelia burgdorferi brain infection" F1000Research, https://f1000research.com/posters/5-127, accessed on August 9, 2020.

27. Kris Newby, The Secret History of Lyme disease and Biological Weapons, Harper Collins, 2019

28. Chartres Cathedral History, https://chartrescathedral.net/chartres-cathedral-history/,Accessed on June 12, 2020

29. Daniel Geary, "Environmental Movement," Encyclopedia.com, last modified May 16, 2020.

Chapter 13: Spiral Dance

1. Jones, S. (2009). "At the still point": T. S. Eliot, Dance, and Modernism. Dance Research Journal, 41(2), 31-51. doi:10.1017/S0149767700000644

2. While the pumping of the heart supports the flow of blood, it is not the main cause of this flow. This fact has been proven by filming blood flow in an early embryo that had not yet grown a heart. See J.L. Bremer, "The presence and influence of two spiral streams in the heart of the chick embryo," *American Journal of Anatomy* 49, no. 3 (January 1932): 409-440, https://doi.org/10.1002/aja.1000490305.

3. Olof Alexandersson, *Living Water: Viktor Schauberger and the Secrets of Natural Energy*, trans. Kit Zweigbergk (Dublin, Ireland: Gill and Macmillan Ltd., 2002).

4. For example, the average combustible engine of a car loses 80 percent of its thermal energy, leaving only 20 percent of the remaining energy to propel the vehicle forward. The majority is expelled as exhaust (which pollutes the environment). Besides being incredibly energy inefficient, it is very expensive since 80 percent of the energy created by petrol is literally wasted in the clumsy workings of the combustion engine.

5. Alexandersson, *Living Water*, 14.

6. William J. Larsen, *Human Embryology*, 3rd ed. (London: Churchill Livingstone, 2001), 315-328, 335-342.

7. Albert Alhadeff, "Michelangelo and the Early Rodin," *The Art Bulletin* 45, no. 4 (December 1963): 363-367.

8. From our vantage point, some 25,000 light years from the galactic center, the Sun speeds around in an ellipse, making a complete revolution once every 220–250 million years or so. (See Ethan Siegel, "Our Motion Through Space Isn't A Vortex, But Something Far More Interesting," *Forbes*, August 30, 2018, https://www.forbes.com/sites/startswithabang/2018/08/30/our-motion-through-space-isnt-a-vortex-but-something-far-more-interesting/#37872f1c7ec2.)

9. "Earth's Motions," Physical Geography, OER services, accessed June 5, 2020, https://courses.lumenlearning.com/suny-geophysical/chapter/earths-motions/.

10. The Editors of Encyclopedia Britannica, "Yuga," Encyclopedia Britannica, last modified November 8, 2015, https://www.britannica.com/topic/yuga.

Chapter 14: Aluna

1. Private meeting and interview with the author Kogi, Northern California 2015

2. *From the Heart of the World: The Elder Brother's Warning*, directed by Alan Ereira (1990; London: BBC). The Kogis had their own indigenous film crew, making it possible for the film to include previously unrecorded holy sites and practices that other film crews had been prohibited from seeing.

3. *Aluna*, directed by Alan Ereira (2012; London: Sunstone Films).

4. Reddy, "What Colombia's Kogi people can teach us about the environment."

5. According to the Vikings.

Chapter 15: Plants: Earth Healers

1. Gautama Buddha, as quoted on GoodReads.com, accessed May 25, 2020, https://www.goodreads.com/quotes/148391-the-forest-is-a-peculiar-organism-of-unlimited-kindness-and.

2. Howard E. Gardener, Frames of Mind: The Theory of Multiple Intelligences, Basic Books 2011.

3. Eric Windhorst and Allison Williams, "Growing Up, Naturally: The Mental Health Legacy of Early Nature Affiliation," *Ecopsychology* 7, no. 3 (September 28, 2015): 115-125.

4. Matthew P. White et al., "Spending at least 120 minutes a week in nature is associated with good health and wellbeing," *Scientific Reports* 9, no. 7730 (2019).

5. Jolanda Maas et al., "Green space, urbanity, and health: how strong is the relation?" *Journal of Epidemiology and Community Health* 60, no. 7 (July 2006): 587-592.

6. Yoshifumi Miyazaki et al., "Combined Effect of Walking and Forest Environment on Salivary Cortisol Concentration," *Public Health* 12 (December 2019).

7. *Lieberman, Gerald A., Hoody, Linda L., "Closing the Achievement Gap: Using the Environment as an Integrating Context for Learning" Published by the State Education and Environment Roundtable, California, 1998.*

8. Michael Klesius, "The Big Bloom—How Flowering Plants Changed the World," *National Geographic Magazine*, July 9, 2020.

9. J.N. Holland and J.L. Bronstein, "Mutualism," in *Encyclopedia of Ecology*, eds. S.E. Jorgensen and Brian D. Fath (Amsterdam: Elsevier, 2008), 2485-2491.

10. David Biello, "The Origin of Oxygen in Earth's Atmosphere," *Scientific American*, August 19, 2009.

11. Anne Marie Helmenstine, "Chemical Composition of the Human Body," ThoughtCo, last modified December 2, 2019, https://www.thoughtco.com/chemical-composition-of-the-human-body-603995.

12. Gavin Buxton, *Alternative Energy Technologies: An Introduction with Computer Simulations* (Boca Raton, Florida: CRC Press, 2015),133.

13. Elisabetta Collini et al., "Coherently wired light-harvesting in photosynthetic marine algae at ambient temperature," *Nature* 463 (February 2010): 644-647.

14. David Biello, "Shining a Light on Plants' Quantum Secret to Boost Photosynthesis," *Scientific American*, February 3, 2010.

15. Mark Fischetti, "Thanks to Plants, We Will Never Find a Planet Like Earth," *Scientific American*, February 1, 2012.

16. John M. Quinn et al., "Land use effects on habitat, water quality, periphyton, and benthic invertebrates in Waikato, New Zealand, hill-country streams," *New Zealand Journal of Marine and Freshwater Research* 31, no. 5 (1997): 579-597.

17. "Importance of Riparian Buffers," West Virginia Department of Environmental Protection, accessed July 15, 2020.

18. United States Environmental Protection Agency Office of Water, "Wastewater Technology Fact Sheet: The Living Machine," Washington D.C.: United States Environmental Protection Agency, 2002.

19. Jennifer Huizen, "Can the Fern That Cooled the Planet Do It Again," *Scientific American*, July 15, 2014.

20. James I. Drever, "The effect of land plants on weathering rates of silicate minerals," *Geochimica et Chosmochimica Acta* 58, no. 10 (May 1994): 2325-2332.

21. Fred Pearce, "Rivers in the Sky: How Deforestation Is Affecting Global Water Cycles," *Yale Environment 360*, July 24, 2018.

22. Fred Pearce, "Earth's most important rivers are in the sky—and they're drying up," *NewScientist*, October 30, 2019.

23. Institute for Systems Biology, "Stress test to predict how diatoms will react to ocean acidification," *ScienceDaily*, June 13, 2018.

24. Maureen D. Keller, "Dimethyl Sulfide Production and Marine Phytoplankton: The Importance of Species Composition and Cell Size," *Biological Oceanography* 6 (1989): 375-382.

25. Intergovernmental Panel on Climate Change, "Summary for Policymakers."

26. Llima Loomis, Trees in the Amazon make their own rain, Science Magazine, Aug. 4, 2010.

27. Mikael Ehn et al., "A large source of low-volatility secondary organic aerosol," *Nature* 506 (2014): 476-479.

28. Dwayne Brown and Alan Buis, "NASA Pinpoints Cause of Earth's Recent Record Carbon Dioxide Spike," nasa.gov, October 12, 2017.

Chapter 16: Thinking Like a Wild Woman

1. As quoted on GoodReads.com, accessed May 27, 2020, https://www.goodreads.com/quotes/857542-a-nation-is-not-conquered-until-the-hearts-of-its.

2. WEA's website is https://womensearthalliance.org/.

3. Interview with the author.

4. International Council of Thirteen Indigenous Mothers, "Alliance Statement," grandmotherscouncil.org, October 13, 2004, https://www.grandmotherscouncil.org/alliance-statement/.

5. Oral story modified by author.

6. This interpretation of the "Through the Eye of the Needle Story" is uniquely my own.

7. Eve Ball, *In The Days of Victorio: Recollections of a Warm Springs Apache* (Tucson, Arizona: The University of Arizona Press, 1970), 15.

8. Peter Charles Hoffer, *Law and People in Colonial America*, 2nd ed. (Baltimore, Maryland: Johns Hopkins University Press, 2019), 71-74.

9. Danny Lewis, "Police Spray Dakota Access Pipeline Protesters With Water and Tear Gas in Freezing Temperatures," *Smithsosian Magazine*, November 21, 2016.

10. Julia Carrie Wong, "Dakota Access pipeline protester 'may lose her arm' after police standoff," *The Guardian*, November 22, 2016.

11. Kring's film, *End of the Line: The Women of Standing Rock*, is currently in post-production. A teaser of the film is available on YouTube: https://www.youtube.com/watch?v=SCxZepDj78A.

12. Cami Anderson, "Why Do Women make Such Good Leaders During COVID-19?" *Forbes*, April 19, 2020.

13. Anderson, "Why Do Women make Such Good Leaders During COVID-19?"

14. James Flynn, Paul Slovic, and C.K. Mertz, "Gender, Race, and Perception of Environmental Health Risks," *Risk Analysis: An International Journal* 14, no. 6 (December 1994): 1101-1108.

15. Noel Bakhtian, "Women's marches, occurring across seven continents, include a focus on environment," *GRIST*, January 19, 2017. Mary Bowerman, "There's even a Women's March in Antarctica," *USA Today*, January 21, 2017.

Chapter 17: Spirit Horse

1. William Shakespeare, *Henry V*, ed. T.W. Craik (New York: Arden Shakespeare, 1995), Act 3, Scene 7.

2. Graham Robb, *The Discovery of Middle Earth: Mapping the Lost World of the Celts* (New York: W.W. Norton and Company, 2013), E-book, unpaginated.

3. Robb, *The Discovery of Middle Earth*, unpaginated.

4. Julius Caesar, "The Siege of Alesia," in *Roman Literature in Translation*, eds. George Howe and Gustave Adolphus Harrer (New York and London: Harper and Brothers Publishers, 1924), 120.

5. A lucid dream is occurring when the dreamer in REM sleep, is aware that he or she is dreaming. Although not a common phenomenon, about 55 percent of people interviewed have stated that they experienced at least one lucid dream in their lifetime. (See Kirsten Nunez, "Lucid Dreaming: Controlling the Storyline of Your Dreams," *healthline*, June 17, 2019.

Chapter 18: Mind over Matter

1. James Jeans, The Mysterious Universe. Cambridge University Press. 1931 Edition. Pg. 137

2. Kalli Szczepanski, "Bushido: The Ancient Code of the Samurai Warrior," September 6, 2019.

3. Gregory Clark and Tatsuya Ishii, "Social Mobility in Japan, 1868-2012: The Surprising Persistence of the Samurai," UC Davis, unpublished manuscript, 2012.

4. Sensei Hidy Hiraoka PhD, "Dr. Hiraoka and Willie show Cosmopower self defense," Youtube Video, February 23, 2015, https://www.youtube.com/watch?v=YfKVeAFANjo.

5. Caroline Lustenberger et al., "Functional role of frontal alpha oscillations in creativity," *Cortex* 67 (June 2015): 74-82, https://doi.org/10.1016/j.cortex.2015.03.012. Matthew D. Sacchet, "Attention Drives Synchronization of Alpha and Beta Rhythms between Right Inferior Frontal and Primary Sensory Neocortex," *JNeurosci: The Journal of Neuroscience* 35, no. 5 (February 2015): 2074-2082.

6. Paramahansa Yogananda, *Autobiography of a Yogi* (New York: Philosophical Library, 1946).

7. Philip Zaleski and HarperSanFrancisco, "100 Best Spiritual Books of the Twentieth Century," Spirituality and Practice, accessed July 13, 2020.

8. Laurie Segall, "Steve Jobs' last gift," *CNN Business*, September 10, 2013.

9. Sybil Lauren, "Peter Hurkos: biography," 2016, 1, https://docplayer.net/14325718-Peter-hurkos-biography.html.

10. Lauren, "Peter Hurkos: biography," 1.

11. V.D. Rusov et al., "Can Resonant Oscillations of the Earth Ionosphere Influence the Human Brain Biorhythm?" arXiv:1208.4970 (August 2012): 2.

12. Mirsolaw Kozlowski and Janina Marciak-Kozlowska, "Resonance and Brain Waves: A Quantum Description," *NeuroQuantology: An Interdisciplinary Journal of Neuroscience and Quantum Physics* 13, no. 2 (2015): 196-204, https://doi.org/10.14704/nq.2015.13.2.795.

13. Amanda Gachot, "Understanding the brainwaves of your children," Up All Hours, accessed July 13, 2020, https://upallhours.com/article/understanding-the-brainwaves-of-your-children.

14. Kavindra Kumar Kesari, Ashok Agarwal, and Ralf Henkel, "Radiations and male fertility," *Reproductive Biology and Endocrinology* 16, no.1 (2018): 118. The International Agency for Research on Cancer (IARC) classified the radiation emitted from cell phones as "possibly carcinogenic to humans."

15. Hagai Levine et al., "Temporal trends in sperm count: a systematic review and meta-regression analysis," *Human Reproduction Update* 23, no. 6 (November-December 2017): 646-659.

16. In 1973 the average Western man had a sperm concentration of 99 million per milliliter but by 2011, that number had drastically fallen to only 47.1 million. This is especially alarming because sperm concentrations below 40 million are considered abnormal. While there is no evidence conclusively proving that harmful EMFs are causing this result, it cannot be discounted.

17. Sascha Segan, "What Is 5G?" *PC Magazine*, April 6, 2020.

18. Sam Roe, "We tested popular cellphones for radiofrequency radiation. Now the FCC is investigating," *Chicago Tribune*, August 21, 2019.

19. Lisa Henkes, "Radio Frequency Radiation (EMF) Threatens Plant and Animal Species with Extinction," *Global Research*, June 11, 2019.

Chapter 19: Nature Quest

1. Mara Freeman, Kindling the Celtic Spirit: Ancient Traditions to Illumine Your Life Through the Seasons (New York: HarperOne 2000), 138. See also, her website at www.chalicecentre.net.

2. David Biello, "The Origin of Oxygen in Earth's Atmosphere," *Scientific American*, August 19, 2009.

3. The Editors of Encyclopaedia Britannica, "Bodhi tree," Encyclopedia Britannica, Encyclopaedia Britannica, Inc, last modified July 10, 2019, https://www.britannica.com/plant/Bo-tree.

4. Sarah A. Laird, "Trees, Forests and Sacred Groves," in *The Overstory Book: Cultivating Connections with Trees*, 2nd ed., ed. Craig R. Elevitch (Holualoa, Hawaii: Permanent Agriculture Resources, 2004), 30.

5. "Irish Traditions: The Celtic Tree of Life," Irish Traditions Online, July 24, 2016, https://irishtraditionsonline.com/celtic-tree-of-life/.

6. "The Prose Tales in the Rennes Didsenchas," ed. Whitley Stokes, *Revue Celtique* 15 (1894): 430.

7. "The Prose Tales in the Rennes Didsenchas," ed. Stokes, 420.

8. John Bradley et al., "Yumbulyumbulmantha ki-Awarawu (All Kinds of Things from Country): Yanyuwa Ethnobiological Classification,"

Aboriginal and Torress Strait Islander Studies Unit Research Report Series 6 (Brisbane: Aboriginal and Torres Strait Islander Studies Unit, University of Queensland, 2006), 1.

9. Rainer Maria Rilke, Duino Eligies, translated by J.B. Leishman & Stephen Spender. The Norton Library, 1939

10. Laura Calabrese, "Yoga and mindfulness practices found to assist breast cancer recovery at cellular level, Canadian study finds," *National Post*, November 3, 2014, https://nationalpost.com/health/yoga-and-mindfulness-practices-found-to-assist-breast-cancer-recovery-at-cellular-level-canadian-study-finds.

11. John Haeglin et al., "Effects of Group Practice of the Transcendental Meditation Program on Preventing Violent Crime in Washington, D.C.: Results of the National Demonstration Project, June-July 1993," *Social Indicators Research* 47 (June 1999): 153, 2011.

12. What is the Super Radiance effect?" World Peace Group, accessed July 5, 2020, https://www.worldpeacegroup.org/super_radiance.html.

Chapter 20: Aether

1. See, for example, the many stone balls from these areas now held in the British Museum's collection: https://www.britishmuseum.org/collection.

2. Plato, *Timaeus*, trans. B. Jowett , retrieved from https://www.gutenberg.org/files/1572/1572-h/1572-h.htm.

3. Philip Ball, *The Elements: A Very Short Introduction* (Oxford, England: Oxford University Press, 2002), 33.

4. Russell M. Lawson, "Elements," in *Science in the Ancient World: An Encyclopedia* (Santa Barbara, California; Denver, Colorado; Oxford, England: ABC-CLIO, 2004), 72.

5. Dr Francesca Minen, A Gust from the East to the West. Warburg Blog, June 2019.https://warburg.blogs.sas.ac.uk/2019/06/12/gust-from-east-to-west-mesopotamian-ideas/

6. Hippocrates, *Hippocratic writings*, ed. G.E.R. Lloyd, trans. J. Chadwick and W.N. Mann (New York: Penguin, 1983), 262.

7. Rachel Nall, "Are there any health benefits to a cold shower?" *MedicalNewsToday*, July 11, 2019.

8. "Celtic Beltane," *Celtic Life International Magazine*, May 1, 2018.

9. "Celtic Beltane," *Celtic Life International Magazine*.

10. Meryl Davids Landau, "This Breathing Exercise Can Calm You Down in a Few Minutes," *VICE*, March 15, 2018.

11. Yogananda, *Autobiography of a Yogi*.

12. A. Pablo Iannone, *Dictionary of World Philosophy* (Oxfordshire, England: Taylor and Francis, 2001), 30.

13. Edmund Taylor Whittaker, "The Theory of the Aether in the Seventeenth Century," in *A History of the Theories of Aether and Electricity from the Age of Descartes to the Close of the Nineteenth Century* (London: Longmans, Green, and Co., 1910), 1-28.

14. Nikola Tesla, as quoted in Giulio Prisco, "Man's Greatest Achievement: Nikola Tesla on Akashic Engineering and the Future of Humanity," Institute for Ethics and Emerging Technologies, December 11, 2015.

15. Paul LaViolette, "The Cosmic Ether: Introduction to Subquantum Kinetics," *Physics Procedia* 38 (2012): 326-349.

16. Tewari's website is https://www.tewari.org/.

17. Paramahamsa Tewari, "Structural relation between the vacuum space and the electron," *Physics Essays* 31, no.1 (2018): 108-129.

18. Malavika Murali, "Bengaluru innovator creates super high-efficiency machine that produces power from vacuum," *Economic Times*, April 7, 2015.

19. High Efficiency Space Power Generator, Invented by Parmanhansa Tewari, The Patents Act, Section 10, 1970, accessed on 11.4.2020 from https://depalma.pairsite.com/Tewari/Tpatent.html & Paramahansa Tewari, "Discovering Universal Reality", published by Tewari.org . Accessed on 11.4.2020 at (https://www.tewari.org/store/p5/Discovering_Universal_Reality_%28PDF%29.html

Chapter 21: Ring of Brogdar

1. Common saying: "He who would be a leader, let him be a bridge."

2. Sigurd Towrie, "The Ring of Brodgar, Stennes," Orkneyjar: The Heritage of the Orkeny Islands, accessed July 15, 2020.

3. Roff Smith, "Before Stonehenge," *National Geographic Magazine*, August 2014.

4. Smith, "Before Stonehenge."

5. Smith, "Before Stonehenge."

6. "Continuation of the Scottish Parliament which at this date made the following Act 'anent the ClanGregour," in *History of the Clan Gregor: Volume First: A.D. 878-1625*, ed. Amelia Georgiana Murray MacGregor (Edinburgh: William Brown, 1898), 436.

7. "The clan battle that led to a ban on MacGregors," *The Scotsman*, February 7, 2018.

8. Stewart Borland, "Clan MacGregor: The History of the Tartan, Crest and Myths," Highland Titles: The Everlasting Gift of Scottish Land, last modified March 18, Highland Titles Limited, 2020.

9. Thomas Smibert, *The Clans of the Highlands of Scotland: Being an Account of Their Annals, Separately and Collectively, with Delineations of Their Tartans, and Family Arms* (Edinburgh: J. Hogg, 1850), 134-135.

10. William Skene, *Celtic Scotland: A History of Ancient Scotland, Volume III: Land and People* (Edinburgh: David Douglas, 1880), 363.

11. Ian Johnston, "The truth about the Picts," *The Independent*, August 6, 2008.

12. David Bates, *William the Conqueror* (New Haven, Connecticut: Yale University Press, 2016), 107-109.

13. John Walker, "Edward Longshanks and William Wallace at Falkirk," Warfare History Network, Sovereign Media, accessed July 15, 2020.

14. George Way and Romilly Squire, *Collins Scottish Clan and Family Encyclopedia* (New York: HarperCollins, 1994), 220-221.

15. The Editors of Encyclopaedia Britannica, "Rob Roy," Encyclopaedia Britannica, Encyclopaedia Britannica, Inc, last modified,

December 24, 2019, https://www.britannica.com/topic/crime-law/Civil-law.

16. The Editors of Encyclopaedia Britannica, "Rob Roy."

17. Jack Boyd, "Stories of the Clan MacGregor and of Rob Roy MacGregor," in *The Road: Reflections on Scottish History* (Bahai Library Online, 2005), E-book, https://bahai-library.com/boyd_the_road#roy.

18. Way and Squire, *Collins Scottish Clan and Family Encyclopedia*, 220-221.

19. Stuart McHardy, *A New History of the Picts* (Edinburgh: Luath Press Limited, 2020).

20. Robert Krulwich, "How Human Beings Almost Vanished From Earth in 70,000 B.C.," *NPR*, October 22, 2012.

21. Michael Greshko, "These Ancient Humans Thrived During the Toba Supervolcano," *National Geographic Magazine*, March 12, 2018.

22. Douglas L. T. Rohde, "On the Common Ancestors of All Living Humans," Massachusetts Institute of Technology, November 11, 2003.

Chapter 22: Cloud Computing

1. Stephen Hawking, quoted in Rory Cellan-Jones, "Stephen Hawking warns artificial intelligence could end mankind," *BBC News*, December 2, 2014, https://www.bbc.com/news/technology-30290540.

2. Elias Aboujaoude et al., "Potential markers for problematic internet use: a telephone survey of 2,513 adults," *CNS Spectrums* 11, no. 10 (October 2006): 750-755, https://doi.org/10.1017/S1092852900014875.

3. Martin Fackler, "In Korea, a Boot Camp Cure for Web Obsession," *The New York Times*, November 18, 2007, https://www.nytimes.com/2007/11/18/technology/18rehab.html?pagewanted=all.

4. Celia Chen, "Inside China's battle to keep internet addiction in check," *South China Morning Post*, June 27, 2019, https://www.scmp.com/tech/policy/article/3016183/inside-chinas-battle-keep-internet-addiction-under-check. Tom Phillips, "'Electronic heroin': China's boot camps get tough on internet addicts," *The Guardian*, August 28, 2017,

https://www.theguardian.com/world/2017/aug/28/electronic-heroin-china-boot-camps-internet-addicts.

5. Maria Nikolaidou, Danae Stanton Fraser, and Neal Hinvest, "Physiological markers of biased decision-making in problematic Internet users," *Journal of Behavioral Addictions* 5, no. 3 (September 2016): 510-517, https://doi.org/10.1556/2006.5.2016.052.

6. American Geophysical Union, "Climate change is making night-shining clouds more visible," July 2, 2018, Phys.org, Science X Network, https://phys.org/news/2018-07-climate-night-shining-clouds-visible.html.

7. The Montreal Protocol, United Nations, Environment Programme Accessed on 10.20.2020 at https://www.unenvironment.org/ozonaction/who-we-are/about-montreal-protocol. Montreal Protocol on Substances that Deplete the Ozone Layer is the landmark multilateral environmental agreement that regulates the production and consumption of nearly 100 man-made chemicals referred to as ozone depleting substances (ODS). When released to the atmosphere, those chemicals damage the stratospheric ozone layer, Earth's protective shield that protects humans and the environment from harmful levels of ultraviolet radiation from the sun. Adopted on 15 September 1987, the Protocol is to date the only UN treaty ever that has been ratified every country on Earth—all 197 UN Member States.

Chapter 23: Younger Brothers & Sisters

1. Catriona MacGregor, *Partnering with Nature: The Wild Path to Reconnecting with the Earth* (New York: Atria; Hillsboro Oregon: Beyond Words, 2010).

2. Private Meeting by the author with the Kogi, Northern California, 2014

3. Wen Bo, China, Simon Elegant, Time Magazine October 2, 2006, http://content.time.com/time/magazine/article/0,9171,1541358,00.html

Chapter 24: The Vision

1. Kahlil Gibran, *The Prophet*, England, Knopf, 1923, pg. 13

2. Bob Sundstrom and Mary McCann, "Around the World, the Soothing Sounds of Birdsong Are Used as Therapy," *BirdNote*, August 19, 2019, The National Audubon Society, podcast transcript, https://www.audubon.org/news/around-world-soothing-sounds-birdsong-are-used-therapy.

3. Doses of Neighborhood Nature: The Benefits for Mental Health of Living with Nature, Daniel T. C. Cox, Danielle F. Shanahan, Hannah L. Hudson, Kate E. Plummer, Gavin M. Siriwardena, Richard A. Fuller, Karen Anderson, Steven Hancock, Kevin J. Gaston, BioScience, Volume 67, Issue 2, February 2017, Pages 147–155, accessed on November 14, 2020 at https://doi.org/10.1093/biosci/biw173

4. Ritesh Ghosh et al., "Exposure to Sound Vibrations Lead to Transcriptomic, Proteomic and Hormonal Changes in Arabidopsis," *Scientific Reports* 6 (2016).

5. Percy Blysshe Shelley, "To a Skylark," Poetry Foundation, accessed May 29, 2020, https://www.poetryfoundation.org/poems/45146/to-a-skylark.

Chapter 25: Fourth Generation Monarch

1. "Constitution of the Iroquiois Nations," ed. Glenn Welker, Indigenous Peoples Literature, February 8, 1996, http://www.indigenouspeople.net/iroqcon.htm.

2. Frank Delaney, The Celts, Grafton Books, a division of Collins Publishing Group, London, 1986.

3. Johanna Paunger and Thomas Pope, The Code: Unlocking the Ancient Power of Your Birthday, Beyond Words, 2011

4. Kate Aronoff, "How Greta Thunberg's Lone Strike Against Climate Change Became a Global Movement," Rolling Stone, March 5, 2019, https://www.rollingstone.com/politics/politics-features/greta-thunberg-fridays-for-future-climate-change-800675/.

5. German Lopez, "It's official: March for Our Lives was one of the biggest youth protests since the Vietnam War," Vox, March 26, 2018, https://www.vox.com/policy-and-politics/2018/3/26/17160646/march-for-our-lives-crowd-size-count.

6. Carolyn Kormann, "The Right to a Stable Climate is the Constitutional Question of the Twenty-First Century," The New Yorker, June 15, 2019, https://www.newyorker.com/news/daily-comment/the-right-to-a-stable-climate-is-the-constitutional-question-of-the-twenty-first-century.

7. United States Department of Defense, "Statement by the Department of Defense on the Release of Historical Navy Videos," Arlington, Virginia: United States Department of Defense, 2020, press release, https://www.defense.gov/Newsroom/Releases/Release/Article/2165713/statement-by-the-department-of-defense-on-the-release-of-historical-navy-videos/.

8. Swami Sri Yukteswar, The Holy Science, 8th ed. (Los Angeles: Self-Realization Fellowship, 1990).

9. Ovid, "Book the First," trans. John Dryden, in Ovid's Metamorphoses, ed. Sir Samuel Garth (London: J. and R. Tonson and S. Draper, 1751), unpaginated, retrieved from http://classics.mit.edu/Ovid/metam.1.first.html.

10. Sri Yukteswar, The Holy Science.

11. "The Ancient Solar Yoga," American Institute of Vedic Studies, The American Institute of Vedic Studies, accessed July 14, 2020, https://www.vedanet.com/the-ancient-solar-yoga/.

12. "The Milky Way's Supermassive Core Sgr A*," The Starburst Foundation, 2010, https://starburstfound.org/mother-star-gravity-well/.

13. Ray Villard and Daniel Stolte, "600 Trillion Suns Light up the Dawn of the Universe," January 9, 2019, The University of Arizona, https://uanews.arizona.edu/story/600-trillion-suns-light-dawn-universe.

14. Karen Oberhauser, "Monarch Butterflies," Journey North: Tracking migrations and seasons, University of Wisconsin-Madison Arboretum, accessed July 16, 2020, https://journeynorth.org/monarchs/resources/article/facts-monarch-butterfly-characteristics.

15. Janet Marinelli, "To Protect Monarch Butterfly, A Plan to Save the Sacred Firs," YaleEnvironment360, Yale University School of the Environment, December 21, 2015, https://e360.yale.edu/features/to_protect_monarch_butterfly_a_plan_to_save_the_sacred_firs.

16. Chip Taylor, "The Big Chill: January 2002," in 2001 Season Summary (Kansas: Monarch Watch, 2001), 9, https://monarchwatch.org/read/seasum.htm.

17. Abigail Derby Lewis, "Following the Monarch Butterflies to Mexico," the Field Museum Blog, the Field Museum, April 8, 2019, https://www.fieldmuseum.org/blog/following-monarch-butterflies-mexico#:~:text=Monarchs%20and%20the%20%22magic%20towns%22&text=For%20centuries%2C%20the%20indigenous%20Pur%C3%A9pecha,time%20to%20harvest%20the%20corn.

18. Sharon Peregrine Johnson, "Butterfly Lore," Lake Waco Wetlands, Baylor University, accessed July 16, 2020, https://www.baylor.edu/lakewaco_wetlands/index.php?id=34628#:~:text=Aztec%20%2D%20associate%20the%20morning%20star,them%20that%20all%20was%20well.

19. Marinelli, "To Protect Monarch Butterfly, A Plan to Save the Sacred Firs."

20. Marinelli, "To Protect Monarch Butterfly, A Plan to Save the Sacred Firs."

21. Kathy Keatley Garvey, "Saving the Monarchs: Saving the Sacred Trees," University of California Agriculture and Natural Resources, December 23, 2015, http://cesanjoaquin.ucdavis.edu/?blogpost=19830&blogasset=13682.

Chapter 26: Upon Landing: Good Works

1. 1 Corinthians 3:8 (NIV), Bible Hub, accessed May 29, 2020, https://biblehub.com/1_corinthians/3-8.htm.

2. Bumble Bees, Levitation and Earth's Magnetic Grid, Talk by Ralph Ring on March 2014, Video accessed on August 2020) at https://www.youtube.com/watch?v=9JrtTtpo7TA&feature=youtu.be&fbclid=IwAR2lAVxHZxWuLElKphFcNn18UCV6lE5_en5dGFYkGc92hjFx--ZNF28d4vs

3. Arthur C. Aho. Tomorrows Energy Need Not Be Fuel. Aldene Books, 1979.

4. Grebennikov, Viktor Stepanovich. "Chapter 5: The Natural Phenomena of AntiGravitation and Invisibility in Insects due to the

Grebennikov Cavity Structure Effect (CSE)." *My World*. Translation posted on KeelyNet.com.

5. Christopher M. Heckscher, A Neartic-Neotropical Songbird;s Nesting Phenology and Clutch Size are Predicators of Accumulated Cyclone Energy, Scientific Reports, July 2 2018.

Chapter 27: Into The Wilds: What Coyote Teaches

1. Samuel Taylor Coleridge, "The Rime of the Ancient Mariner (text of 1834)," Poetry Foundation, accessed March 7, 2020, https://www.poetryfoundation.org/poems/43997/the-rime-of-the-ancient-mariner-text-of-1834.

2. National Academies of Sciences, Engineering, and Medicine, Evaluating the Taxonomic Status of the Mexican Red Wolf and the Red Wolf (Washington, D.C.: The National Academies Press, 2019), https://doi.org/10.17226/25351.

3. Amando Cockrell, "When Coyote Leaves the Res: Incarnations of the Trickster from Wile E. to Le Guin," Journal of the Fantastic in the Arts 10, no. 1 (Winter 1998): 64-76, https://www.jstor.org/stable/43308325.

4. "Coyote Makes for Unusual Guest at Chicago Quiznos Restaurant," BusinessWire, April 3, 2007, https://www.businesswire.com/news/home/20070403006289/en/Coyote-Unusual-Guest-Chicago-Quiznos-Restaurant.

Chapter 28: Becoming a Runaway Horse

1. Andrew Cohen, "The Feds Unnecessarily Round Up Wild Horses, Then Complain About Costs," The Atlantic, May 2, 2012, https://www.theatlantic.com/national/archive/2012/05/the-feds-unnecessarily-round-up-wild-horses-then-complain-about-costs/256527/.

2. Center for Biological Diversity, "Grazing," Center for Biological Diversity, The Center for Biological Diversity, accessed July 16, 2020, https://www.biologicaldiversity.org/programs/public_lands/grazing/index.html#:~:text=Livestock%20grazing%20is%20promoted%2C%20protected,may%20be%20three%20times%20that.

3. Center for Biological Diversity, "Grazing."

4. Rachael Bale, "This Government Program's Job is to Kill Wildlife," National Geographic, February 12, 2016, https://www.nationalgeographic.com/news/2016/02/160212-Wildlife-Services-predator-control-livestock-trapping-hunting/.

5. Vickery Eckhoff, The Real Price and Consequences of Livestock Grazing on America's Public Lands

Report analyzes taxpayer bailout of U.S. public lands ranching [Part II of a series on ranchers], Western Watersheds Project, February 12, 2015. https://www.westernwatersheds.org/sustainable-cowboys-welfare-ranchers-american-west/Accessed on August 29, 2020

6. Ibid

7. See Clark Wissler, "The Influence of the Horse in the Development of Plains Culture," American Anthropologist 16 (1914): 10.

8. Soshi Parks, The Shared History of the Wild Horses and Indigenous People, Yes Magazine, April 27, 2020. https://www.yesmagazine.org/environment/2020/04/27/native-horses-indigenous-history/ accessed on July 29, 2020

9. Ann Forsten, 1992. Mitochondrial-DNA timetable and the evolution of Equus: Comparison of molecular and paleontological evidence. Ann. Zool. Fennici 28: 301-309.

10. Feist, J. D., & McCullough, D. R. (1976). Behavior patterns and communication in feral horses. Zeitschrift für Tierpsychologie, 41(4), 337–371. https://doi.org/10.1111/j.1439-0310.1976.tb00947.x

11. Yvette Running Horse Collin ,THE RELATIONSHIP BETWEEN THE INDIGENOUS PEOPLES OF THE AMERICAS AND THE HORSE: DECONSTRUCTING A EUROCENTRIC MYTH, Doctoral Dissertation University of Fairbanks Alaska, May 2017, downloaded from https://www.sacredwaysanctuary.org/publications on August 21, 2020

12. Wendell Berry, "The Peace of Wild Things," Scottish Poetry Library.com, accessed May 26, 2020, https://www.scottishpoetrylibrary.org.uk/poem/peace-wild-things-0/.